© 1970 by
PRENTICE-HALL, INC.
Englewood Cliffs, New Jersey

All rights reserved. No part of this book may be reproduced in any way or by any means, without permission in writing from the publisher.

Current printing (last digit):

10 9 8 7 6 5 4 3 2 1

13-578906-0
Library of Congress Catalog Card Number: 74-114007
Printed in the United States of America

METHODS AND MATERIALS OF CONSTRUCTION
A Guide for Builders, Owners, Architects, and Engineers

LAURENCE E. REINER

Real Estate Consultant, Darien, Conn.
Retired Vice President, Equitable Life Assurance Society

Prentice-Hall, Inc., Englewood Cliffs, New Jersey

METHODS AND MATERIALS OF CONSTRUCTION
A Guide for Builders, Owners, Architects, and Engineers

PRENTICE-HALL INTERNATIONAL, INC., *London*
PRENTICE-HALL OF AUSTRALIA, PTY. LTD., *Sydney*
PRENTICE-HALL OF CANADA LTD., *Toronto*
PRENTICE-HALL OF INDIA PRIVATE LTD., *New Delhi*
PRENTICE-HALL OF JAPAN, INC., *Tokyo*

PREFACE

The field of construction is such a vast one, I have found it necessary to limit myself to certain aspects, particularly as they relate to the proper use of materials in buildings and the methods used to install or erect them. I have also emphasized that buildings provide shelter for people at home, at work or at play. The architect engineer must constantly bear this in mind. His design must provide the amenities as well as the necessities.

I have not attempted to teach design but have confined myself to the economic, physical and aesthetic complexities of construction. My reader's attention is called to the wealth of excellent material available in handbooks and textbooks on design of which he can avail himself for further study.

Above all, I have tried to emphasize the importance of good design, good construction and good materials. This knowledge is a distillation of 42 years of experience in every area of the building business and I have attempted to show that "man need not live by shelter alone" but can have beauty inexpensively. A handsome, well-built structure has a lifetime rental and ownership appeal.

My gratitude to my friends and colleagues in this field who have given me the run of their reference files and have kindly read my manuscript—to my faithful Swanee who typed it—to my proofreader who made it readable—and to Karin, my wife, who sat patiently through many long evenings while this book was being written.

LAURENCE E. REINER

ACKNOWLEDGMENT

The author wishes to thank many of his friends and associates who were of great help in furnishing data and some of whom were kind enough to check several of the chapters for accuracy in light of the latest advances in the technology of construction.

Some but not all were:

Burton Bailen
Roy O. Allen
Albert Kennerly
Robert Ebner
Robert Crimmins
Max Abramovitz
Emanuel Pisetzner
John J. Somers
E. J. Berlet
John Kemp
Robert Sullivan

CONTENTS

1. Site Selection and Site Planning 3
2. Zoning and Building Codes 13
3. Preparation of Plans and Specifications 25
4. Contracts and the Bidding Process 39
5. Excavation and Foundations 51
6. The Structural Frame 73
7. The Structural Floor 101
8. The Electrical Installation 111
9. Heating, Ventilation and Air Conditioning 125
10. Plumbing and Sprinklers 161
11. Vertical Transportation 175
12. The Exterior Wall 195
13. Floors and Ceilings 223
14. Sound Conditioning 237
15. Stairs, Miscellaneous Iron and Roofing and Flashing 247
16. Interior Masonry, Lathing, Plastering, Painting and Decorating 265
17. Ceramic Tile, Metal Toilet Enclosures and Accessories 277
18. Ornamental and Hollow Metal Work 283
19. Hardware 287
20. Finishing the Interior for Occupancy 291

CONTENTS

21 LANDSCAPING 299
22 CABINET WORK AND MILLWORK 305
23 SCHEDULING A CONSTRUCTION PROJECT 311
24 DESIGNING FOR THE HANDICAPPED 321
25 SAFETY OF THE PUBLIC, WORKERS, AND PROPERTY DURING CONSTRUCTION 325
26 AT THE COMPLETION OF THE CONSTRUCTION, HOW SOME FUTURE PROBLEMS MAY BE AVOIDED 335
INDEX 341

METHODS AND MATERIALS OF CONSTRUCTION

A Guide for Builders, Owners, Architects, and Engineers

CONTENTS

1. SITE SELECTION
    A. The Private Residence
    B. The Multifamily Dwelling
    C. Special-Purpose Buildings
        1. The manufacturing plant
        2. The laboratory and the medical building
    D. The Office Building
        1. The suburban small office building
        2. The major office building

2. SITE PLANNING
    A. The Private Residence and Subdivision
    B. The Multifamily Structure
    C. The Manufacturing Plant
    D. The Research and Medical Building
    E. The Office Building

1

SITE SELECTION AND SITE PLANNING

1. SITE SELECTION

The proper selection of the site for any construction project is of great importance and can make the difference between success and mediocrity. The knowledgeable owner will call in his architect engineer before deciding upon the site, and the architect should be familiar with the basic requirements of a successful location.

1A. THE PRIVATE RESIDENCE

Let's start with a private residence. The architect engineer has been asked by the owner to look at one or more sites which have been tentatively selected. Before he does so, however, the architect engineer should make it his business to know the owner; his likes and dislikes; the size and ages of his family; their hobbies; their interests. With these in mind, he sets out to satisfy his client.

First he investigates the zoning, to find out whether the location is in a restricted residential zone; what size plots are allowed; how far the site is from an unrestricted or lower class zone; what the limitations are on use, area, and height. He should investigate how strictly the Zoning Code is enforced. Some Boards are more lenient than others in granting variances, often

SITE SELECTION AND SITE PLANNING

to the discomfiture of the neighborhood. He also notes the surrounding houses and the state of their upkeep to determine whether the area is stable.

Second, he looks into the features of the location that ensure enjoyable living. If the owner has children, the distance from schools and the major or busy highways to be crossed are important considerations. The location of bus lines or the commuter railroad and the shopping areas should be observed. If there is a likelihood of ice or snow, the access roads to the location should be considered.

Third, he looks at the site itself and its topography. Trees may add a new and interesting dimension, as can the architecture of the surrounding houses. He should look for rock outcrops which may necessitate expensive foundations or low spots which may produce wet basements. An intimate knowledge of the site will help him create a better house.

It may seem that after considering and evaluating all the very evident things that all should be well. The experienced architect, however, must consider other important matters such as the direction of the sun and the prevailing wind. The ingenious designer will plan his residence to profit from the sun's light and warmth. If the prevailing wind is from the southwest and so is the town *dump,* extreme caution should be exercised and such considerations could even preclude the site.

1B. THE MULTIFAMILY DWELLING

In choosing locations for other than private homes, the architect engineer team is usually brought in before a site is purchased. The owner or client is presumed to be knowledgeable, but the planner must be depended on to have a working knowledge of the zoning and building codes in the area chosen. For the multifamily dwelling or apartment house there are many things to look for before the architect engineer approves the site. First is the Zoning Code. Such codes not only state which are the approved residential areas but set forth such things as height and land coverage restrictions. Here again it is important to make a thorough study to see that nearby areas are not zoned for objectionable uses. Second, the planner must re-examine such matters as access to public transportation, highways, schools, and shopping. It is true that the owner has done this, but he will welcome the planner's re-examination. Third is a careful examination of the site itself. If the owner of the land will permit, it is well to sink a few test borings to see what underground problems may be encountered. The shape of the plot and the direction in which it faces may help the planner decide among several choices. For instance, a long narrow plot with the short dimension facing the street does not give the owner much opportunity to provide balconies, which bring premium rentals. If he provides balconies on the long dimension, subsequent high-rise buildings may cut off the view. If the site is on high land it may provide sweeping views but may be difficult to get to in icy weather. The successful planner must think of these matters.

1C. SPECIAL-PURPOSE BUILDINGS

The location of sites for special-purpose buildings such as manufacturing plants, laboratory or medical use, or large or small office buildings, require that the architect engineer as well as the owner have a thorough understanding of the exact use intended for the building. They should know where the work force is coming from; what utilities are required; what the zoning is.

1C1. *The Manufacturing Plant*

Let us examine the requirements in order. The modern manufacturing plant is usually on one level. It must be accessible to major roads and/or a railroad. Because the zoned location for such plants is in outlying areas and away from residential areas, the work force usually comes by automobile. There must be extensive facilities for unloading raw material and loading finished material. It therefore requires a considerable amount of land area. It is essential that tentative layouts of the amount of land required be made before any site is acquired, keeping in mind possible future expansion needs.

Attention must also be given to the availability of the utilities required for the particular manufacturing process. There may be heavy power requirements, and in such a case large power lines should be within reasonable distance. If a great amount of water is required and there are no public water lines nearby, one or more wells will be needed, and prior investigation must be made of the quality and quantity of water available. Waste water must be disposed of through drain fields or by piping to public sewer lines or to nearby streams. If the plant uses heavy equipment, potential foundation problems should be investigated.

1C2. *The Laboratory and the Medical Building*

There are hundreds of types of special-purpose buildings but we shall limit ourselves to two: the laboratory building and the medical building. The laboratory or research building is constantly becoming more common. In many suburban communities there may be several of these buildings, devoted to research on anything from air and water pollution to the best way of baking bread or the investigation of human reactions. Such communities are restudying their ordinances in order to allow such use, but have set very definite requirements. The architect engineer must carefully study the code.

SITE SELECTION AND SITE PLANNING

He may find such requirements as hidden parking, extensive planting, or a review of his architecture by a town board. The client may be allowed to build in the area but it may not be the building he wants. Very often the more burdensome requirements may be removed by a reasonable appeal to the town boards involved. Each of these structures has its own peculiar site requirements. One should be away from noisy traffic; another may require extensive grounds for privacy. The laboratory may require process water or a large power supply. Extensive parking facilities are required. Such buildings can be quite handsome and set in extensive landscaped surroundings.

The planner should play a major part in the *medical building* site selection. His clients are usually not well informed about zoning codes or parking requirements or height restrictions. The planner must look for a site that provides easy parking for patients and easy access to the proposed building. The author knows of a case where the parking facilities for a medical building were between columns at the top of a steep ramp. The patient had to climb a flight of stairs to get to the elevator lobby. This building was occupied by an orthopedist and the author had a cracked kneecap! A knowledgeable planner would have warned the owner against the purchase of this site or advised him to grade it to furnish a proper entrance for future patients.

1D. THE OFFICE BUILDING

1D1. *The Suburban Small Office Building*

Every town of any size now contains at least one new office building, and the growth in the number and size of such comparatively small structures is without parallel before the 1960s. Here, of course, zoning and area and height restrictions are of paramount importance. Most towns require parking areas in proportion to office space; many towns require planting areas; there are likely to be rather severe restrictions on height so that more land area is required for a given building. The intelligent architect engineer will make himself familiar with all the governing rules and will advise the owner as to the site best suited to his needs.

1D2. *The Major Office Building*

In the case of the major office building in a large community the planner is always called in to advise the owner or client. Very often an option is taken on a site until the highest and best use of the land can be determined. During such an option period the architect engineer must study the Zoning Code for the area and try various combinations of structural shapes and sizes that come within this code in order to determine whether the client's needs can be met. He must also investigate the possible foundation conditions by having test borings made. If the requirements cannot be met within the existing code, it is often possible to obtain a variance from existing provisions or even to have an area rezoned to a higher use. Such an effort takes expertise on the part of the planner in preparing necessary drawings and in marshalling arguments (in combination with an attorney) as to why such a variance or rezoning should be granted. A property which is rezoned to a higher use can more than double its value. Grateful owners make excellent clients.

2. SITE PLANNING

A number of examples of site selection for various uses have been mentioned. With the site now chosen, the rest of this chapter will be devoted to examples of how various-use structures should be located on the land. It is assumed that in most cases the architect engineer has helped select the land and has certainly considered how the proposed structure would fit.

2A. The Private Residence and Subdivision

In the case of a private residence, it should be located, if possible, with regard to land elevations, to existing trees, and to the direction of the rising and setting sun. The ingenious planner can use a slope to plan a multi-level house or can use existing trees to shade or frame his structure. The portions of the residence that profit most from the sun's light and warmth can be located accordingly. In the 1960s where land areas permit, development houses have been placed in a circle or arc with a portion of the land area allotted to each house and with most of the land devoted to a community recreational park. In this way children are kept off the streets and each house has a view.

2B. The Multifamily Structure

To properly locate the multifamily, or apartment house, the sun directions and the proximity of noisy streets or highways must be considered. The planner should also consider the character and bulk of the nearby structures.

Zoning restrictions regarding density of population and area and height restrictions must be considered. Some regulations set forth minimum distances between buildings. All zoning ordinances have requirements for off-street parking and many cover such matters as size and location of planting areas. In orienting the building, the bedrooms should be away from noise; the living room should have sun and should face east or west if it can't face

SITE SELECTION AND SITE PLANNING

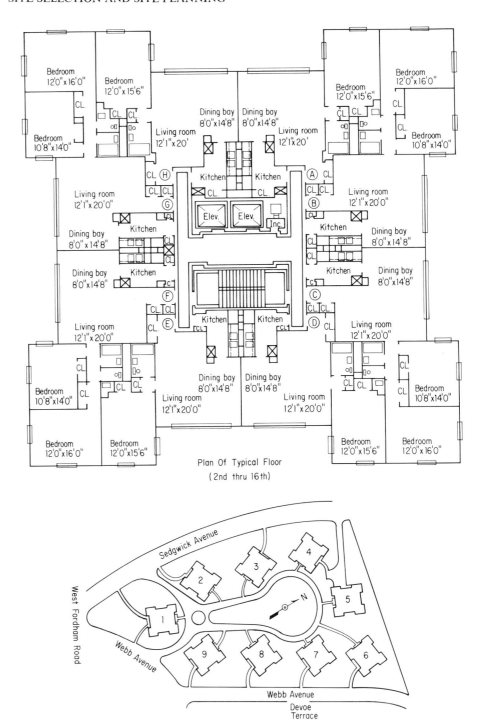

Plan Of Typical Floor (2nd thru 16th)

Fordham Hill

University Heights, New York City. The West Looks Out Over the Harlem River Valley and Upper Manhattan. The East Looks Over a Park.

Parklabrea, Los Angeles. Metropolitan Life Insurance Company.

south. The balcony or living room should have the best view. Obviously not every apartment can have all these advantages but they may be provided for in the larger, more expensive units. Practical architecture? Yes. The individual buildings of garden apartments should be located so that a tenant cannot look directly into the window of another tenant. A plan that takes advantage of rock outcrops or existing large trees can make a very appealing setting. Shown on these pages are two examples of site planning. The *first figure* shows a large multifamily development on high land (Fordham Hill), commanding sweeping views. It can be seen that the buildings are on the perimeter of the land area. They are located so that even the inward looking units have a view of the interior gardens and a glimpse at the large view between the opposite buildings. They do not face each other squarely so that an appearance of privacy is preserved. The *second figure* shows a garden apartment project (Parklabrea) on flatland. The only views are those the planner provides. By use of wide landscaped roads and walks and skillful planning, the apartments have a vista and yet still retain a sense of privacy. Both of these projects are highly successful.

SITE SELECTION AND SITE PLANNING

2C. The Manufacturing Plant

The manufacturing plant must be located with practical considerations foremost in mind but should still be well designed and landscaped. The location of the main highways with respect to the site is a prime consideration. The parking lots must be arranged conveniently with respect to these highways and to the plant. Lots marked to correspond to the plant sections that are nearest to them are obviously helpful. If possible, there should be many entrances and exits—also, if possible, in the event of bad weather, the entrances to the plant should not be located too far from the lots. The location of a railroad siding with relation to loading docks or the maneuvering area for trailer trucks must be studied. It is necessary to provide storage areas for flammable materials both raw and finished. The skillful location of these facilities with the most efficient relation to each other helps to create a successful plant.

2D. The Research and Medical Building

The laboratory, research, or medical building is often located in a residential zone and, as noted earlier in this chapter, the planner must pay careful attention to the restrictions. He can provide convenient parking and skillfully planned vistas and for the medical building he can provide convenient access, gentle grades, and privacy.

Gateway Center, Pittsburgh, Pa.

SITE SELECTION AND SITE PLANNING

Rockefeller Center, Inc., New York, N.Y.

2E. THE OFFICE BUILDING

The location of an office building on a site, unless it is an office building complex, is almost determined by the plot itself. Usually the land is expensive and the owner requires maximum coverage within the Zoning Code and the area and height restrictions. The structure should face a principal street. The planner can take advantage of alternate provisions in some codes to beautify his structure by skillfully placing it on the plot. In Chapter 2 "Zoning and Building Codes," this will be explained.

The planner has much more scope in the site planning of an office building complex. There are several notable examples of such planning. In the two shown here the eye is directed along landscaped vistas. The buildings in Gateway Center in Pittsburgh all take advantage of the view and are set at angles to each other. At Rockefeller Center the low buildings have their long axis parallel to the landscaped channel gardens and act as a frame for the 60-story high RCA building. This building seems to soar even higher due to the vista of the gardens and the sunken skating rink which draw the eye down before it reaches this structure. By now the reader must realize how much the proper location and skillful planning of a site add to the aesthetic appeal, the convenience and the financial success of the project.

CONTENTS

1. ZONING REGULATIONS OR CODES
   A. Their General Purpose
   B. The Separation of Various Land Uses
      1. The suburban town and the small city
      2. The large city
      3. The planned development
   C. Area and Height Restrictions
      1. In residential areas
      2. In business areas
      3. In transitional areas
         a. The possibility of a variance

2. BUILDING CODES
   A. Their General Purposes
   B. The Basic Requirements of Any Code
   C. 1. The small community
      2. The city—the additional requirements
         a. Protection against fire
         b. Structural requirements
         c. Health and safety regulations

2

ZONING AND BUILDING CODES

In the preceding chapter on site location and planning, the subject of zoning and building codes was frequently mentioned. The architect engineer, or owner builder, must be constantly aware of these codes and regulations because everything he plans is subject to them.

1. ZONING REGULATIONS OR CODES

1A. Their General Purpose

The general purpose of a Zoning Code is to provide for the orderly growth of a community and to promote the general health, welfare and safety of the public. A well drawn code will control the density of population; it will prohibit the use of land for purposes detrimental to a neighborhood; it will, by designating certain areas for business, residential, or industrial use, encourage healthy growth; and at the same time will avoid transportation snarls and school overcrowding. It can even encourage good architecture. The Regional Plan Commissions in metropolitan areas can be considered master zoning boards and by their planning promote the safety and well being of an entire region—sometimes comprising portions of several states and containing millions of people. Such plans show major highways, railroads, and airports, and study air and water pollution and rubbish disposal. The intelligent architect engineer has respect for such codes and makes every effort to work within them.

ZONING AND BUILDING CODES

1B. THE SEPARATION OF VARIOUS LAND USES

1B1. *The Suburban Town and the Small City*

A typical suburban town may separate its land uses as follows: separate residential areas in which a single family home may be built on a land area ranging from one-fifth of an acre to four acres; apartment house areas which will usually be allowed only in the lowest category of residential uses or in retail business areas; such use is limited to a minimum required plot "area" as for example: "no plot less than 10,000 square feet and not more than 10 units per acre." It also describes business uses which allow small office buildings, retail shops, restaurants, and small theatres.

The small city code is somewhat more liberal than that of the suburban, or "bedroom," town. The city is a self-contained unit and must allow land to be used for all purposes necessary to sustain it as a viable community. To quote from such a code: "Certain uses with established functions in the economy but having a well-known nuisance potential are permitted only in the Industry District and there only by special exception." Nuisances are allowed but under severe restrictions. The suburban town does not allow them at all. The small city may also zone industrial land to attract business and industry.

The small city code also allows smaller building lots in residential areas than suburban towns do because land is more expensive and all utilities are usually in place. Small service shopping centers in certain selected parts of residential areas may be allowed, as well as professional offices and multi-family apartments. Its requirements for density of population may allow 20 housing units per acre instead of the 10 allowed in a suburban town. In special central areas where utilities are in place and good transportation, shopping, and other amenities are available, there could be as many as 70 housing units per acre. This permits a fairly good-sized apartment house.

1B2. *The Large City*

The zoning ordinance of a big city presents an interesting study for the architect engineer. We are all aware of the problems of big cities—bad housing; insufficient recreational areas; poor transportation to industry from slum areas; crowded schools and many others. City planning can go a long way through proper long term zoning, toward correcting some of these ills. Such forward planning should be of interest to anyone who is ever going to build anything. There is a great need for the involvement of the architect engineer and owner builder. Intelligent planning in improving our cities can be greatly rewarding.

During the past few years several great cities have completely revised their zoning to meet the challenge of the times—the need for the orderly upward revision of the uses of the land. Some of the purposes of such a revised code are as follows:

ZONING AND BUILDING CODES

To promote the health, safety, morals, and welfare of the people.

To protect the character and maintain the stability of the various use areas.

To regulate the intensity of the use of the various zoned areas.

To prohibit uses which are incompatible with the various use areas.

To provide for the gradual elimination of those present uses which do not conform to the standards of their area and which are adversely affecting it.[1]

To provide for the condemnation of such buildings or land when it is found necessary for the rehabilitation of the areas blighted by such buildings or land uses.

The last two purposes are at the heart of the matter. They provide the authorities with a weapon to combat the growth of the slum and to rehabilitate such areas by the right of "eminent domain" or condemnation. Again, because land is expensive and utilities and streets are in place and public transportation is available, greater density is allowed in all areas. The highest graded residential areas will allow seven houses to an acre and in the lower graded residential areas a multifamily dwelling can be built to house as many as 50 families per acre.

A city zoning map will show how the city planners have tried to combine large homogeneous areas, and by setting aside small areas for local shopping and small businesses have shielded such areas from adverse use. Such maps show a gradual shift from residential to local business to general business to local commercial, general commercial, and finally industrial.

The big city must allow all sorts of nuisance uses which create noise and noxious odors and which include the manufacture and use of corrosives and explosives. Such uses, however, are strictly regulated for noise and air and water contamination and such plants must keep their distance from safe and quiet industrial operations.

1B3. *The Planned Development*

In all cities big and small and even in some surburban towns the codes have begun to allow "Planned Developments." This is of particular interest to the architect engineer and owner builder. If a sufficiently large tract of land can be acquired (the area of the allowable tract depends on the use area) then a good deal of freedom is given to the planner to develop a residential, business, or industrial park. Height and area regulations are not as restrictive and he can create super blocks and interesting combinations of structures instead of abiding by the usual city gridiron pattern.

1C. AREA AND HEIGHT RESTRICTIONS

The second purpose of the zoning codes, after regulating the use of land, is to provide for its orderly use within each established area. In all zones, it therefore sets height and land coverage and bulk restrictions. It

---

[1]Some cities allow a nonconforming use to remain for 15 years before it must be moved.

ZONING AND BUILDING CODES

states how far from the lot lines one must build, how much off-street parking is necessary and how many family units or how much office area may be built on the land. The area and height restrictions must be studied carefully by the planner. Essentially they tell him what kind of structure he can plan for his client.

1C1. *In Residential Areas*

In the surburban town, in an effort to keep the density of population down and to provide a semirural atmosphere, restrictions are severe, especially in outlying areas where there may be no utilities in place and no public transportation. Multifamily dwellings, if not entirely forbidden, are severely restricted as to land coverage and height.

In the cities, even in outlying residential areas, a more intensive use of the land is permitted. The most restricted use of land in a city is about the same as the highest density allowed in a suburban town. The same comparative restrictions hold true for multifamily apartment dwellings. While a suburban town may restrict the apartment house to two stories or a height of 30 feet and to 30% of the lot area, the small city may allow apartment buildings of four stories with a 30% lot coverage in its lower grades of residential areas and in its high density areas it may allow a building of 10 stories if the building is situated a sufficient distance from the lot lines. It can be readily seen how such provisions provide for light and air and reasonably low density.

In the big city where land costs are high and public transportation is readily available the code will allow as many as 400 dwelling units per acre, compared to the 10 families per acre in the suburbs. All cities allow certain bonuses in the way of extra height for a building that faces a wide boulevard or sets back from its permissible building line. The idea is to always provide maximum light and air and minimum density within economic limits.

1C2. *In Business Areas*

Business buildings in suburban towns usually cannot be over three stories or 40 feet high. The business building can occupy almost all of its land except for minimal side and rear yards but it must provide off-street parking immediately adjacent to it. Most small towns also allow research and laboratory buildings in special areas, which as stated in Section 1B3 of this chapter (entitled "The Planned Development"), presents a challenge and opportunity for the architect engineer to do something outstanding for his client and to add to the character and beauty of the environment.

The small city usually attempts to draw a line between restrictions which will make it uneconomical to build a business building and rewards for the design of a building that will allow maximum light and air to its surroundings. It may allow a gross building area of twice the lot area in outlying business districts and six times the lot area in downtown areas. It doesn't usually mention height or lot coverage as such because obviously the total lot area and maximum permitted building area determine this. It may state (for instance) that for every foot of front yard provided back of the lot line the building gross area may be increased 1% above the allowed base amount. This could mean that on a 10,000-square foot lot where a 60,000-square ample of rezoning an entire large area. This is a wide main thoroughfare which verted and to regulate the use, occupancy and maintenance of these buildings or solid plaster interior partitions.

The big city business regulation must allow for more intensive use of the land in the central areas. In outlying areas the ratio of lot to building gross area may vary from as little as from 1.2 to 4. In or near the central business district the ratios can run up to seven times the lot area and in the central business district the ratio of gross building area to lot area for an office building may be as high as 16 times. The regulation may state that if the building lot faces a public space at least 200 feet wide then the building to lot area ratio may be increased by 15%. The regulation may continue by offering bonuses[2] for setbacks from the building lines. That is why one can build the 50-story-high Marina City apartments or the 100-story-high John Hancock office building in Chicago or the 110-story-high World Trade Center in New York City. The same regulations apply to a lesser extent for industrial zones.

All city zoning regulations require off-street loading facilities. Some cities require off-street parking. Some large cities discourage central city parking facilities. They wish to keep private cars from choking the streets and hindering the free flow of necessary commercial traffic and public transportation. This is especially true where public transportation is good.

1C3. *In Transitional Areas*

The possibility of a variance in all communities large or small, the pattern of growth and the change in the character of certain areas must be recognized by the zoning authorities. In a small town if the residential character of the areas immediately adjacent to the central district degenerates, the intelligent authority will consider an appeal to grant a variance for a business use or to rezone. Such an appeal must be presented by the owner and his planner to show the authority that it will add aesthetic character to the area. The increased value of the land and the increased taxes for the town are never mentioned.

In the big city the authority will be sympathetic to business uses in or near blighted residential areas, especially if such use is meant to employ residents of the immediate area. Wilshire Boulevard in Los Angeles is an example of rezoning an entire large area. This is a wide main thoroughfare which

---

[2] All Zoning Codes offer inducements to the owner and his architect engineer to keep their structures away from the property lines. The owner can build a higher building and obtain more rentable space and the city gets large airy plazas and wide streets.

leads from the central city west for 20 miles and taps residential areas for nearly its entire length. It is a natural site for intensive business use and the authorities have recognized this. However the application for change to business use must show the city fathers the benefits to the city. The architect engineer should lead the team that presents the appeal. He must show by the character of his plan that it will benefit the city and the people.

2. BUILDING CODES

2A. Their General Purposes

The purpose of any building code is to provide minimum standards and requirements for safe design, methods of construction, and uses of material in all buildings or structures newly erected, altered, repaired, or converted and to regulate the use, occupancy and maintenance of these buildings or structures. As in the Zoning Code the building code encourages good architecture and engineering. The architect engineer *must* be familiar with the general provisions of the local codes. In some large cities where such building regulations are quite complicated there are local architects and engineers who make a career of interpreting the code and acting as intermediaries between the owner, the architect engineer and the local building authority. Such people are known as "retained consultants" and are recognized in every architect-engineer-owner contract. The intelligent designer will not, however, depend on these consultants entirely, and in small communities he should have completely independent knowledge.

2B. The Basic Requirements of any Code

Many suburban towns and smaller cities use a statewide code and adapt some of its provisions to meet their peculiar needs. Some states publish two codes: one a basic code for guidance in all communities and the other an abridged code which is for the use of smaller towns which are residential in character and allow only low business buildings. Some states have adopted the uniform code as approved by the Building Official Conference of America (the BOCA code); the Federal Housing Administration has prepared minimum requirement building codes for residences which supplement the state codes.

Following are the basic requirements of a building code. Section 2C2 of this chapter will present the additional requirements of the large city code. Because the single family residence is the most basic of structures the code requirements start with it and become more stringent as they apply to larger structures where more people live and work.

Fire Resistance. The nonfireproof frame residence which is set back from all lot lines and street lines need only comply with rules for fire stopping to prevent the spread of fire through draft openings. An attached garage must be isolated by fire retardant plaster and a solid door.

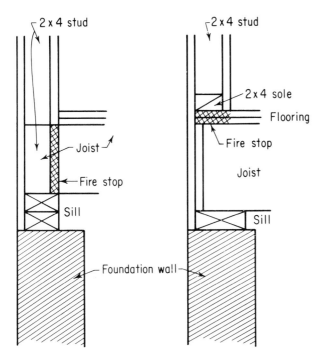

The Cross-Hatched Lumber Is The Fire Stop That Prevents A Fire From Spreading Up Between The Vertical Studs

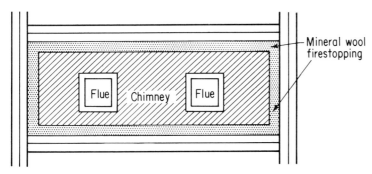

Framing And Firestopping Around A Chimney

Typical firestopping.

All regulations cite certain fire resistive qualities around furnace chimneys or fireplace flues and around the furnaces themselves. Larger private residence may have to be of noncombustible construction to qualify for the maximum allowable floor area. Noncombustible materials mean masonry exterior walls, light steel interior framing and floor beams and masonry or solid plaster interior partitions.

Structural Strength. The requirements start with the footings and foundation walls. The depth of footing from the grade or finished ground level depends on the climate. In northern states such depth is usually 3′6″ but can go to 4′0″ or 4′6″. The minimum width and thickness of the concrete footing and the foundation walls is specified. The allowable floor and roof loading is given (usually 40 pounds per square foot for floors and 30 pounds per square foot of horizontal projection for roofs except where heavy snow load-

19

ZONING AND BUILDING CODES

ing on roofs may increase this). The minimum sizes of joists, columns, supporting beams and rafters are given for various clear spans and loads. The type and sizes of framing, bracing, and sheathing are given.

Sanitary and Heating Facilities. The code for sanitary plumbing because of its health implications is usually statewide. It goes into such details as the sizes and materials of piping for various purposes. The materials and design of plumbing fixtures are specified. In smaller towns where there may be no sewer lines or public water in outlying areas it mentions in detail the sizes and types of septic tanks and drain fields and the location of wells, and periodic testing of the water supply. The requirements for heating plants usually confine themselves to safety and fire resistance rather than to capacity. They will specify such things as furnace and flue cut out switches, pressure relief valves and approved combustion devices. As mentioned under "Fire Resistance" the code will also specify minimum distances between hot surfaces and combustible material such as walls, floors, or ceilings.

Light and Ventilation. This section provides for minimum amounts of both natural light and air and for minimum habitable room sizes.

Electrical. All small towns and many cities both large and small use the National Electrical Code as their guide. The basic state codes may modify the requirements to suit local conditions and the large cities more often than not write their own codes, but essentially the National Code sets minimum standards everywhere. Because the improper installation of electrical work presents such a hazard to the safety of people and property it is strictly regulated. The size and material of every main feeder line, meter board, branch circuit, panel board, circuit breaker, etc. is rigidly set for every type of use and every capacity. The protection of all wiring and all electrical devices is set forth in detail. In any type of structure the code always calls for the installer to have a license or the equivalent to perform such work. The planner should also be familiar with the National Fire Underwriters' Code.

2C1. *The Small Community*

The requirements for the multifamily building in a small community are of course more restrictive than that for a single family residence, but because the Zoning Code limits such structures in area and height, the code differs from that for the private residence mostly in the fire resistance and general safety requirements. It requires two means of egress; fire resistive walls around fireproof stairways; it specifies the distance to the nearest means of exit; it severely limits or forbids the construction of mutiple dwellings of entire wood frame construction. The structural strength is of course set by its size and height. In the sanitary plumbing code it will not allow a multiple dwelling where there is no town sewer. It assures sufficient water and sanitary facilities for each unit. It becomes stricter in the sizing, installation, and protection of electrical circuiting and an underwriter's certificate is almost always required.

The surburban or small town business building, as is the multiple dwelling, is limited in size and height by the Zoning Code. The local code usually defers to a more stringent state code because there are likely to be more people involved than in a residential structure. A business building is allowed more area per lot because it is in a comparatively crowded central business district. The code insists on masonry or metal exterior walls and fire retardant if not fireproof construction. It calls for 50 pounds live floor loading for a minimum design. It requires two widely separated means of egress and fireproof stairways. The plumbing requirements specify the number of toilet facilities that must be installed. The code specifies the maximum legal occupancy which is usually one person per 100 square feet. The stairway widths are based on such occupancy. All codes allow additional occupancy through the allowance of extra height and area where a building contains automatic sprinklers.

The type of construction and the facilities of a place of public assembly are also severely restricted. For instance, a nonfireproof theater is not permitted at all and a nonfireproof motion picture theater may be only one story and contain 3500 square feet, whereas a completely fireproof motion picture theater can be as large as 20,000 square feet. The code will limit occupancy and will prescribe multiple exit facilities which, of course, depend on the allowed occupancy. Most places of public assembly are subject to statewide and Fire Underwriters' codes as well as the local one.

2C2. *The City—The Additional Requirements*

The large city building code is very complicated because it must provide for large apartment buildings, large office buildings, large theaters, and industrial complexes of all kinds where thousands of people may be working. The architect engineer who will design the structure must be thoroughly familiar with it. Before he puts pencil to paper he must know how large his building can be; what allowance he must make for stairways and elevators; how large his toilet rooms must be; how much ventilation he must supply, and so on. In combination with the Zoning Code he should know what bonuses he will get for the use of certain materials and installations and in the design of the bulk of the structure involving ratios between heights and areas. When the architect engineer is designing for a plot that may cost as much as several hundred dollars per square foot such extra credits become extremely valuable and do not sacrifice good design.

. . . Because building codes vary to some extent from city to city it is best here to present only the highlights. It must be borne in mind that however they vary, all regulations stress good design and are concerned with health, welfare, and safety.

ZONING AND BUILDING CODES

2C2a. *Protection against fire*

We have seen that even in small communities, where no more than two-or three-story buildings are allowed, that exit requirements and fire resistant qualities of materials are quite strict. Much more attention must be paid to these requirements in multistory buildings where hundreds of people live or work. The code therefore severely limits the height and area of non-fire resistive structures. It classifies fire resistance by types which indicate hourly ratings for various portions of the structure. Such ratings are determined by the number of hours such materials or portions of the structure can continue to perform their function when exposed to great heat and flames. Tests are standardized by The American Society for Testing Materials (ASTM) or by the National Board of Fire Underwriters. Therefore the more fire resistive the structure, the larger and higher it may be *depending always on the use zone where it is to be located.*

Fire protection of the structure is always paired with the exit requirements. A code will require only one exit stairway in a small and low residential or business building. As the structure gets higher and larger the exit requirements are for at least two-hour walls[3] surrounding the stairs and the corridors leading to them, and for at least two separated concrete or steel stairways.

2C2b. *Structural requirements*

The code almost always starts with a recital of the minimum design loads, which are the pounds per square foot of live load[4] that structures for various occupancies must be designed to carry. The design of the structure therefore is based on the building not only holding itself up but also supporting the live load of people, equipment, filing cabinets, machinery, and anything else that an ingenious occupant can think of. The allowable foundation loading is determined by tables giving the maximum weight per square foot that can be placed on almost every possible type of soil, sand, clay, or rock. The materials of the foundation and the methods by which such materials may be placed are carefully specified. All codes call for constant inspection of all work by qualified experts either privately retained or publicly employed. The public authority always has the last word in case of any doubt with respect to the stability or strength of the design. The requirements for structural strength are explicit for the small building whereas for the large multistory building most codes refer to the specifications for design of steel or concrete structures as promulgated by the American Institute of Steel Construction (AISC), the American Concrete Institute (ACI) or the Portland Cement Association (PCA). Such organizations as these make it their

---

[3] A two-hour wall is capable of resisting intense heat and flames for two hours. All codes mention fire resistance by a period of time.

[4] The live load is the weight the structure must carry above its own weight.

ZONING AND BUILDING CODES

business to issue and constantly amend specifications and instructions to provide safely designed and erected structures, and such specifications are enforced by the authorities. As noted previously, while most cities use uniform trade codes as mentioned above (always corrected, of course, for special hazards such as earthquakes or very high winds), they become more explicit and vary widely in their requirements for interior partitioning, elevator shaftways, exterior curtain walls, use of certain floor systems, etc. Here the architect engineer must become familiar with the specific requirements of his own locality.

2C2c. *Health and safety regulations*

Now that fire protective and structural strength requirements have been designed into the structure according to the code, we come to the amenities that make it convenient and enjoyable to live or do business in the structure. However, lest the architect engineer, at the orders of an owner client, tries to skimp—the code requires certain minimum facilities. In multiple residences there must be a sanitary facility with specified requirements for size, light, and ventilation in every unit. The code specifies the kind and number of plumbing fixtures. In the business building the code requires a certain number of basins, water closets, and urinals per unit of population and for each sex. It specifies how much air must be circulated in all kinds of buildings by either calling for minimum areas of side and rear yards or for mechanical ventilation into interior spaces where natural light and air are not available. In industrial buildings the code specifies special shower facilities, drinking fountains, and special ventilation for particular hazards. There are minimum requirements for heating plants and their protection. So far as the electrical installation is concerned most local codes defer to The National Electrical Code which to some extent varies in different communities, but its standards of safety and its specifications for materials and methods are universally respected.

The codes go one step further. They contain certain minimum standards for the protection of workmen and the public during construction. They require the protection of openings in floors by the use of railings; they forbid the use of open fires in the building; they require life lines around the perimeter of open floors; the property must be fenced from the public and sidewalk sheds constructed to protect the public against falling objects. In the hoisting of men or materials the code is very specific about safety devices.

If this chapter has been read carefully, the intelligent architect engineer or owner will note that most new codes and regulations are meant to be of help in attaining acceptable standards of design and construction. They are also meant to protect the legitimate planner and owner against the competition of shoddy materials and workmanship, and structures which without such codes would crowd their environment and possibily endanger life, health, or safety.

CONTENTS

1. UNDERSTANDING THE CLIENT'S NEEDS
    A. The Residence
    B. The Business Building
        1. The office building
        2. The multiresidential building
        3. The special-purpose building
    C. A Reminder and Recommendation to the Planner

2. THE DESIGN
    A. Structural Design
        1. The structural frame
        2. The foundation
        3. The floor system
    B. Electrical Design
    C. Mechanical Design
        1. Air conditioning and heating
        2. Plumbing
    D. Vertical Transportation
    E. Consideration of Future Maintenance

3. THE ARCHITECTURAL CONCEPT
    A. Assembling the Areas Required and the Zoning Requirements into an Envelope
    B. The Placement within the Envelope of the Mechanical Components
    C. The Treatment of the Envelope

4. THE SPECIFICATION

3

PREPARATION OF PLANS AND SPECIFICATIONS

Previously in this book the architect engineer was told to be sure that he understands his client's requirements before planning any structure. The architect engineer uses this understanding in combination with his own knowledge of codes and good building practice and his desire for good design to produce his plan.

1. UNDERSTANDING THE CLIENT'S NEEDS

1A. THE RESIDENCE

Let us now discuss a client's needs by using as an example a private residence. The skillful architect "sizes up" his client and attempts to match his own building concept to the client's personality. He must determine the family composition; mode of living; hobbies requiring special design; a preference for a particular material or architectural notion; the number of rooms required and their uses. In many cases the client has only a hazy idea of what he wants. Then the architect should act as counselor and guide his client toward a mutually acceptable plan.

1B. THE BUSINESS BUILDING

1B1. *The Office Building*

The approach is similar in the case of the small entrepreneur who wishes to build a medical or laboratory building, garden apartments, or a two-story *taxpayer*. (A taxpayer is a two- or three-story building with stores

PREPARATION OF PLANS AND SPECIFICATIONS

on the first floor and offices above.) The skillful architect will respect his client's wishes, while leading him toward the proper decisions.

Most large structures, however, are proposed by experienced and knowledgeable clients who have a good idea of what they want. When such a client wants to build a high-rise apartment or office building, the architect must first determine from the client the kind of occupancy demanded by the prospective tenants. This may consist of large enterprises requiring considerable clerical help and large interior spaces, or small branch offices requiring perimeter offices. The occupancy expected by the owner is often determined by the locality and the local rental situation. A study of newly built buildings is helpful. Sometimes the owner already has one, or even more, major tenants with special requirements. Such requirements can be anything

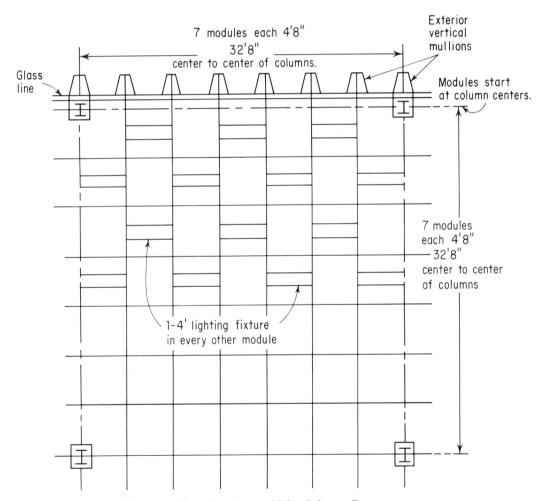

A Typical Modular Plan Showing a 32'8" Column Bay. Structural Steel Framing, Underfloor Ducts, Air Conditioning, Partition Lines, Lighting Fixtures, Are All Adapted to the Module.

PREPARATION OF PLANS AND SPECIFICATIONS

from special dining facilities or executive toilet rooms to interior stairways. They can also result in a change of the entire concept as from a conventional to an "all electric" building.

The architect, with the owner, must also at this time determine and agree upon a module. This is a unit of measurement from which the entire basic structure is evolved. For instance, a five-foot module means that the columns will be arranged so that an even number of five-foot sections of office partition, or five-foot sections of exterior building skin, can be fitted between them. This means, also, that interior space can be arranged so that the ten-foot or two-module office is the minimum enclosed space. If smaller minimum offices are considered, or if an exterior architectural concept is a determining factor, then a 4'6", or a 4'8" module can be considered. (The diagram shows a typical modular plan.) In one major office building, a 3'6" module was used and the minimum office is three modules, or 10'6", while the lighting and partitions are on 3'6" centers. This is unusual, however, and many special (nonstandard) sizes of building products had to be purchased. (If the order is large enough, it doesn't cost any more but generally standard sizes should be used.) From such information a plan is evolved which takes into consideration column spacing; ratio of perimeter to interior space; lighting patterns; the heating, ventilating, and air conditioning system, the vertical transportation, and many other important aspects which make the difference between an ordinary building office and a comfortable business home.

1B2. *The Multiresidential Building*

In planning a high-rise apartment house the problem is somewhat different. Here the need is for comfortable living. No large, unobstructed areas are required. The important considerations are traffic between rooms, the possibility of going to a bedroom without passing through the living room, the availability of the toilet facilities, the wall space for furniture arrangement. Sometimes too little attention is given to sound insulation etc. of floors and walls. Before planning, the architect engineer must determine the requirements of the owner and the rental market. The rental scale that is to be charged, the neighborhood, the views, must all be taken into account. Such things determine the entire concept, such as the size of rooms, the number of bathrooms, the kitchen facilities, the exterior treatment (balconies or not), the arrangement and decoration of the entrances and lobbies, and many other amenities that make for pleasant living within the various income brackets.

1B3. *The Special-Purpose Building*

The medical building, the laboratory, or any special-use building must be the product of the closest cooperation and understanding between the client and the architect engineer. In such cases, however, the owners or

PREPARATION OF PLANS AND SPECIFICATIONS

user's requirements are paramount. The need for such facilities as special piping for compressed air, various gases, corrosive liquids, or high-voltage electricity, must be studied and accommodated within the structure. In a medical building, the physician must be readily accessible to his patients. For example, a study should be made of any special hardware or entrance ramps required. The spacing of offices between treatment rooms or laboratories is important. Special air conditioning or exhaust requirements must be noted. Obviously the architect engineer must investigate and research the matter thoroughly. He should study plans of medical, laboratory, or other special-purpose buildings and trade publications, and obtain literature from trade organizations.

1C. A REMINDER AND RECOMMENDATION TO THE PLANNER

In addition to all these considerations the planner must also give high priority to building costs. The private residence must be within the client's means. It is discouraging to have to redesign a house because the client could not afford the original design or to have to compromise quality of space or material in a large building because bids have been higher than anticipated and the project has become uneconomical. In any project, but especially in the large project, the planner must know his contractors; he must know something of the labor supply and wage situation (in timing his project); he should always keep in mind that he must strive to furnish *the highest quality possible, both practically and aesthetically, for each dollar spent.*

2. THE DESIGN

Having studied and comprehended the client's requirements and the various requirements and regulations for land use, height, use of materials, parking areas, special requirements for hazards, and any other requisites of the duly constituted bodies having jurisdiction, the architect engineer is now ready to start his design. He has compiled all his data and can now start a plan from which he can build a suitable environment.

2A. STRUCTURAL DESIGN

2A1. *The Structural Frame*

One of the first steps in starting a plan is the structural design. There are certain basic things which must be determined. For an office building, one of the first is the column spacing. It is usually important that large bays free of columns be provided. A bay is the space between four columns. The size of the bay is determined by the module, which has been mentioned previously. Such a bay enables the prospective tenant of the building to arrange his space into various smaller working entities without having the columns intrude any more than is necessary. It is important that all buildings be built on a modular system so that within the column spacing, standard interior partitioning or lighting can be installed in even numbers. In the case of a highrise apartment building, (steel or concrete), the column spacing need not be as wide as for the office building and can be made to fit the interior room layout. A living room that is 15′ × 22′ would certainly call for a column bay, and other rooms could be fitted within such a bay size. Column spacing is also important to suit the requirements of the interior core of an office or an apartment building. The core columns must be so spaced that they will allow clear passage for elevator shafts, stairways, ventilation and electrical shafts, and free access to central toilet facilities, and other facilities that are required in a multistoried, multitenanted building. In the case of a manufacturing plant, the column spacing is definitely a product of the manufacturing process and the architect engineer must determine his structural system. There is a structural system which is best suited for each use and the architect engineer must be familiar with these variations.

2A2. *The Foundation*

Before the architect engineer starts his plan, he is, of course, aware of the foundation situation. He is in possession of test boring data and other pertinent information. The architect also knows that his structural designer will be able to design a suitable foundation for almost any structure within reason which allows the architect freedom to devote his time to his overall concept.

2A3. *The Floor System*

The floor system requires more immediate attention because the type of structural floor used does affect the overall design. In developing his plan the architect has been in constant touch with his client and with the various engineering consultants on the design team. During these discussions the preliminary decisions will have been made as to the kind of structural system to be used; the electrical system the building will have; the type of air conditioning to be furnished and how it will be distributed, and other matters affecting the basic design. The structural floor design must fit these various requirements. It is strongly recommended that weekly meetings be held between the owner and the design team during such early design stages and that the minutes of each meeting be kept.

PREPARATION OF PLANS AND SPECIFICATIONS

2B. Electrical Design

At the same time, the interior arrangements and services which make a building a workable environment must be considered and many basic decisions made. Among these is the electrical system. In this day when electricity is being applied to almost every aspect of domestic and business use and comfort, the forward looking architect engineer should not only understand his client's present needs but should incorporate provisions into the structure which will allow for the possible future expansion of the system. He must plan circuiting for lighting, office equipment, both internal and external communications, and data processing. In the present-day building air conditioning and ventilation use a large amount of power. This volume of electricity must be supplied through a system of main feeders, branch feeders to junction boxes, circuit wiring to individual areas; and all of this wiring goes through main circuit breakers, branch circuit breakers, and switches. A big building electrical system is not unlike the human circulatory system with its arteries, veins, and capillaries. For the apartment building, provisions may have to be made for electrical cooking or individual air conditioning. In all homes the increased use of electricity calls for heavier circuit wiring and more and more circuits. Some communities will not grant a building permit unless a service of a certain minimum capacity is installed. In the case of special-use buildings, the client's plans for equipment utilization are the deciding factor. The forward looking planner tries to provide for more capacity than is required by present or planned equipment and for a flexible system whereby equipment can be moved from place to place without the tenant or owner having to install entirely new heavy circuiting.

Sometimes the client may wish to use electricity for everything—heating, cooling, air conditioning, lighting. In this case, the entire concept may be affected. Although the cost of electricity has decreased in recent years and may be quite low in some communities, it is still a comparatively expensive way of heating, and the architect engineer must plan his building to take advantage of every economy. Chapter 9 gives some examples of this, in "all electric" buildings. The planner must also investigate his source of electricity and become familiar with the regulations of the local power company. He must recognize that the electrical installation requires many different kinds of space and must now plan his building to accommodate such spaces. He must provide space for vertical shafts—to take care of heavy electrical risers. He must also provide space for transformer rooms, for meter rooms, for panel boards, for circuit breakers and somewhere in the floor, or above the ceiling, he must provide room for circuit wiring. Some of the heavier portions of an electrical installation can be placed in the basement or on the roof, or in the mid-floor of a building. This all must be thought our carefully in order to provide the most economical space for a well functioning system.

2C. Mechanical Design

2C1. *Air Conditioning and Heating*

In these times no structure, whether it be multiresidential, office, laboratory, medical, or for some other special use, is even considered without seriously considering air conditioning or cooling. The public is becoming more and more conscious of air pollution and along with the ever increasing demand for skilled help to work at tasks requiring concentration, a comfortable all-year climate has become a business necessity. The skilled and knowledgeable planner should be familiar with at least the standard methods, in addition to the many other ways, of providing such an environment. Here again, and this can't be stressed too often, the needs of the client must be consulted. The air conditioning for a multiresidential property can vary widely from that for a major office building or a special-use structure.

Cooling can be accomplished in several ways, such as by means of centrally placed chillers or compressors, by absorption units, by heat pumps or by individual smaller units. Consideration must be given to the placement of large central units. Recently there has been a trend toward such placement on the roof instead of basements of large buildings. The advantages of this and of various basic systems will be discussed in a later chapter.

All cooling systems perform their function through the transmission of air. This is true for every type of air conditioning from the common window unit to the largest office building system. As the building becomes larger and the population within it increases, an enormous amount of air must be moved to furnish a pleasant environment. Air in such large quantities is of measurable weight. The architect engineer must provide for taking such air into the building, cooling it, and sending it by means of giant fans to where it is required. He must send more air to some places than to others. Consideration, therefore, must be given as the plan progresses, to the location of the large equipment which is needed to move such air. Shaft space must be provided for the very large ducts that move the air vertically from the central fan rooms to the various floors of the building, and area has to be provided above finished ceiling or below finished floor, to move the right quantities of air to the places on the floors where it is required.

The heating of a building also uses air as the vehicle. The source of the heat is usually a hot water boiler which may use gas, oil, or electricity as its fuel. The hot water that is generated is sent to the same places that the cold water for summer cooling has gone, and heats the air to the proper temperature. In many structures air has to be both cooled and heated in different places at the same time.

All structures, regardless of size, have to be zoned for heating and cooling. In such cases, account must be taken of the exposures, the exterior surface materials, the prevailing winds, and the distribution of the building population. Where there is an industrial use of air cooling or heating, provisions must be made for special cooling of heat-producing equipment or for extra heating and cooling of areas where large quantities of air must be exhausted for safety or processing reasons. The entire cooling and heating system of a building, therefore, becomes a function which requires large spaces for fans and ducts and compressors, boilers and pumps, provision for large supplies of air, electricity, water, and know-how.

2C2. *Plumbing*

Although the plumbing installation in a conventional-use building is unobtrusive, and everyone takes it for granted, it is not any less important. The planning of readily accessible, comfortable, and attractive sanitary and kitchen facilities is a vital part of any structure in which people will live or work. In a single-family residence the proportion of baths to bedrooms can make or break a sale. In multifamily housing the accessibility and appeal of the bathrooms and kitchens make renting easier. The success of an office building and the comfort of its inhabitants is dependent on the convenience and attractiveness of the toilet and washrooms, in addition to the ready accessibility of cold drinking water, and water for cleaning and other purposes. The same applies to an industrial plant, where the accessibility of attractive toilet and washing facilities are factors that can contribute toward a satisfied labor force.

In the special-purpose building the plumbing system does much more than supply comforts. It becomes a vital part of the purpose for which the entire structure has been built. The medical or laboratory building, with its needs for apparatus that uses all sorts of liquids or gases, special storage tanks or valves or waste tanks, cannot function without the sanitary and process engineer who works along with the architect in studying and supplying these needs. The manufacturing plant, which may require tremendous quantities of water, requires intelligent planning for supplying the water and for disposal of the waste.

Fire protection is an important consideration in many buildings. Most codes call for sprinkler protection in special hazard areas. Paint-storage rooms or underground garages usually must be equipped with sprinklers. All multioccupancy buildings have standpipes which can obtain water from either the street or from a roof tank. The water supply, the sanitary facilities, the disposal of waste, and other facilities requiring piping must be tied in with the other facilities of the building in order to produce a functioning whole.

2D. VERTICAL TRANSPORTATION

The methods of transporting people and materials vertically range from the ladder to the stairway to highly sophisticated elevators and moving stairways. Even on raw construction jobs, if the structure is tall, provisions must be made for mechanically hoisting supplies and men. This is a matter of simple economics. A steel worker or mechanic will be less tired and therefore more efficient if he has been raised by a temporary elevator or construction hoist to the twentieth floor, and today such construction elevators are safe and comfortable. The stairway is the basic means of vertical transportation. When all else fails, the stairs are still there. All building codes specify the requirements for the width and materials of stairways and walls surrounding them. The planner must satisfy such code provisions and allow room for them.

Nowadays it is rare to find a new structure that is over two floors high and has no elevator. Equipment can range from a simple machine for a small structure, to the highly sophisticated, electronically controlled high-speed elevator which almost thinks for itself as it sets up traffic patterns. The number, size, and speed of elevators to be planned for a high-rise building are the result of a calculation that uses areas, population, and other factors. In Chapter 11 a more detailed explanation of this will be given. This must, however, be determined as the building is being planned. The moving stairway, or escalator, has recently become an important consideration in mass vertical transportation. In many large structures such as terminal buildings or department stores, or in some interfloor communication use in office buildings, it has become a necessity. The number of shafts, the number of floors served by each shaft, the location of elevator machine rooms, the location of cross-over floors from one set of shafts to another—all must become part of the master plan.

2E. Consideration of Future Maintenance

We have discussed such basic design components in the building as the structural design, electrical design and mechanical design. In all of these components there are materials and methods of installation that can often be varied when the architect engineer considers maintenance. By starting with more expensive materials, the planner can very often save the owner or user much expense and inconvenience later. For instance, the selection of flooring materials is important, whether it be for industrial plant, apartment or office building, or private residence. There are many choices of floor treatments, from those for heavy trucking to those for top grade institutional buildings, and the final choice should be thoroughly investigated. The planner can choose wall surfaces of many varieties. Sometimes the most expensive and most handsome floor or wall surface is not the best for long-term maintenance. The writer knows of one case where very expensive paving requires constant patching, and of another, where the handsome wall covering can be easily scratched (and is!). The walls of elevator cabs should be durable as well as beautiful.

Ceiling materials vary in their washability and some ceiling surfaces are fragile and dent easily. The material of piping should be determined by the liquid or gas that it carries—but this is not enough. Special hazards must be considered. For example, consider the case of water—water is not just water; it may be "soft" or "hard." It may carry a good deal of dissolved carbon dioxide, or various minerals, or it may be slightly alkaline or acid. The wrong piping may literally dissolve in a very few years. The choice of electrical motors, or heavy compressors, or fans must be determined by the service they are to perform—whether they will be used 24 hours a day or intermittently, once or twice a week or just for standby.

For a large structure, the planner should try to use the products of only one or two manufacturers for all similar mechanical or electrical equipment, so that an interchange of parts is possible. In the case of equipment that runs for long periods of time the planner should be familiar with the durability of the shafting and bearings. He should carefully plan the foundations under such equipment in order to abate sound transmission and vibration. Lighting fixtures should be designed so that lenses and tubes can easily be replaced. In elevator installations, all major companies will recommend a maintenance contract, whereby they will furnish labor and material to do everything from replacing a set screw to recabling or replacing the hoist motors. The consideration of accessibility, durability of materials, and replacement costs are an important part of the planner's job.

3. THE ARCHITECTURAL CONCEPT

3A. Assembling the Areas Required and The Zoning Requirements into an Envelope

From the basic steps of the planning of the structure, the architectural designer member of the architect engineer team has been developing an idea of how the various and necessary components will be gathered together into an harmonious and workable whole. Such planning is necessary, even for the simplest structure. For the private residence or smaller business-use building, the Zoning Code has been studied and the designer knows his client's preferences. The planner should now be able to draw a *building envelope* within which all these area and functional requirements are satisfied. The variation possible in the private residence is much wider than for the small business-use building. In any case, the economy of construction is important and unusual shapes cost money. Within the budget, however, the skillful designer still has a fairly wide latitude. Any building can be made long and narrow or square and tall, as long as its envelope comes within the zoning regulation, and will in its usable areas, satisfy the owner's needs.

The design of the envelope becomes much more complicated for the multitenanted high-rise business building. Here the designer must balance zoning requirements, building codes, and the total enclosed area required, against the budget. According to Zoning Code requirements in some cities, he can only build so high before he has to set back or he can set back at street level and go straight up. His building can have a central core in which the elevators, stairways and other utilities such as electrical and ventilating risers are concentrated, or this core can be at one end or another; it can be a protuberance from the building, or it can even be separated into two parts. To satisfy the owner's preference (which is quite often determined by economic considerations), and depending on the site, the building can be long and narrow or comparatively square. For example, 10,000 square feet of office building area can be placed within a perimeter consisting of a square with each side of 100 feet, or a total of 400 feet, or a perimeter consisting of two sides each of 200 feet and two ends each of 50 feet, and having a total length of 500 feet. This building, which could have its core along one long side, contains a large amount of perimeter space and might be suitable for small office users. It is of course an impractical shape and would be expensive to build.

In the case of the multifamily residential building, the designer must, within the envelope, try to see that bedrooms have cross ventilation; that living rooms have a view; that within each apartment there is easy room-to-room communication; that no bedroom adjoins a neighbor's living room. In satisfying these requirements he could end up with a fairly weird shape. It is sometimes necessary to compromise in order to arrive at a viable design. Even in designing some institutional buildings, where economics are of secondary consideration (the dream of all designers), he must still produce an envelope that is workable within the codes and that will provide for all the necessary uses within it.

3B. THE PLACEMENT WITHIN THE ENVELOPE OF THE
MECHANICAL COMPONENTS

Now that the building envelope has been determined, the architect engineer team must fit within it all the functions that will make it a living entity. This stage of the planning is a pulling, twisting, and squeezing type of operation. The designer and owner want good open rentable space and the engineers want space for running their electrical feeders and their ventilating ducts, and for their fan rooms and compressor rooms, and elevator machine rooms. The floor to floor height must be adjusted to give clear ceiling height and at the same time allow all the electrical circuits, and lighting fixtures, and branch ventilating ducts to fit into the ceiling or floor. The structural design must fit in with all these requirements. The planners must check to see that all vertical shafts are clear and that a plumbing riser doesn't go through the middle of a ventilating duct. The stairways must be planned so that every space on a floor has two legal means of exit. In the industrial building or special-purpose building where large clear space may be required, the power lines, and ventilating and heating lines, and all process piping must be fitted in and around, or over or under these spaces. In the multi-residential building, the lines for the many separate kitchens and bathrooms must be provided for and still not create odd corners in rooms or corridors. This entire fitting together is a result of long and painstaking cooperation within the architect engineer owner and (very often) contractor team. Every one knows what all the others are doing. Even after this, it is the fortunate building operation that does not have dozens of field coordination problems and very often extra orders to correct discrepancies, which mean extra expenditures for the owner.

3C. The Treatment of the Envelope

At this point in the planning, when the size and shape of the structure have been determined and the internal functions are about to be fitted into the building, the architect turns his attention to the external treatment. The entire concept of the structure has, of course, been thought of and sketched, but the final architectural treatment cannot be considered until all its components have been brought together into a practical and workable whole that meets all the requirements of codes and owner. Depending on his budget and to some extent on the owner's preferences, the architectural designer has a considerable amount of freedom. He can vary the external design of a residence from Olde English Tudor to what he thinks is contemporary. He can dress up a store building or taxpayer to make it look Colonial or Georgian. Even industrial buildings can be made pleasing to the eye by the skillful designer's use of good proportions and materials. One-story cement block warehouses can be made attractive by laying the block in a pleasing pattern; by using a little brick or wood trim at the front; by a good entrance design and by skillful use of planting, even if it is only a window box. It costs very little extra and is good business. The exterior treatment of the multistory residential building is very important. The building will be home to many families and its treatment, both the exterior, as to fenestration, balconies, and entrance, and the interior, as to lobby and corridors, should be warm and in a scale that is not monumental, but man-size.

The large office building presents a unique problem. First, the planner must determine how the building is to be used; whether it is to be a status symbol for a large public or institutional enterprise, or whether it is to be a speculative building to be rented for profit. Perhaps it is a monumental building built by a wealthy foundation. The design of a fabricated metal and glass, or masonry skin can be widely varied from almost flat to boldly proportioned. The glass area should be determined. The use of stone, either real or precast, or brickwork, will help to present the image that is desired. The costs vary widely. Where heavier materials are used, the structural frame of the building and its foundations may have to be specially designed for them. The exterior of the building is what the great majority of people see and it should be designed to tell these people its story at a glance.

4. THE SPECIFICATION

We come now to the writing of the specifications. This work has really been going on to some extent during the entire planning, but it does not become final until the plan is almost completed. The specification is the performance guide for the builder. The specification and the plan complement each other. The specification also tells the builder what all the component systems of the structure are expected to do. It spells out the quality of all the thousands of parts to be used. It describes methods by which parts are to be fitted together or installed within the structure. It mentions acceptable manufacturers and always gives several choices of any product. The plan shows where the part is to be installed and the specification describes the part and tells how it is to be installed. All building contracts state that if a valid construction detail is forgotten in either the plans or specifications, the builder is liable for it and must furnish it. The writing of a construction specification requires a skillful painstaking person who has a wide knowledge of the products and methods available to perform certain tasks. Frequently the specification writer thinks of a completely new way of installing a particular part or even of a new system. The specification writer is usually an architect or an engineer, specializing in a certain field and he must, as much as anyone, know how the structure is to be used and how it is expected to perform. If a certain lighting pattern or light intensity is decided upon he must describe how it is to be done by specifying the size of fixture and type of lens. He must describe such things as material testing or how structural members are to be joined, or the construction and material of a valve for a certain duty. After the specification writer has completed his task and the plans and the specifications are checked and rechecked to see that together they make a harmonious whole, then the package can be given to a builder to bid or to build.

CONTENTS

1. THE ARCHITECT ENGINEER'S CONTRACT
   A. The General Provisions
      1. Some types of contracts
   B. Some Duties of the Architect Engineer
      1. Field Inspection
      2. Shop Drawings
      3. Approval of Payments

2. SELECTION OF A BIDDER'S LIST
   A. General Contractors
   B. Subcontractors

3. CONSTRUCTION CONTRACTS
   A. How a Bid is Solicited and Assembled
   B. The Lump Sum
   C. The Cost Plus Fixed or Percentage Fee
   D. The Negotiated Contract
   E. The Guaranteed Maximum Plus Fixed Fee
   F. The Scope Contract
   G. The General Conditions of a Contract
   H. Use of a Cost Estimator

4

CONTRACTS AND THE BIDDING PROCESS

In the first three chapters, the author has often mentioned the importance of the relationship between the architect engineer and owner builder or general contractor. Sometimes, especially in smaller communities, this relationship is simply established by a handshake. However, building anything can become a fairly complicated procedure. People do not always remember exactly what was said and understood between them. To save future embarrassment, monetary loss, and even worse consequences, a simple contract which sets forth an understanding is strongly recommended. No one should feel embarrassed to ask for a written agreement (after all people can die or become incapacitated). As projects become larger, and when the associates who are to form the association necessary to build them, are large business enterprises, then a contract is always necessary.

1. THE ARCHITECT ENGINEER'S CONTRACT

1A. The General Provisions

Usually before any amount of work is done by the planner for an owner or client, except possibly in site selection, an agreement has been reached. The contract between an owner and architect engineer must contain certain basic general provisions. First, it is dated and names the parties that are entering into the agreement. Second, it sets forth the location and what the work is, such as a multistoried office building, or a shopping center, or a private residence. Third, it states what services the architect engineer is to perform—whether he will furnish both architectural and engineering services; whether he will help in the selection of a general contractor and subcontractors; whether he will furnish construction inspection service; whether he will furnish certain retained specialists for such possible items as vertical transportation, kitchen or restaurant layouts, etc. Fourth, the contract will mention the compensation or fee and should state how and when it is to be paid as the planners' work progresses. This portion of the agreement should also mention the reimbursement for out-of-pocket expenses for such things as necessary travel, out-of-town telephones and telegrams, extra blueprinting, and fees for obtaining various building permits and other governmental approvals.

1A1. *Some Types of Contracts*

There are several kinds of contracts that can be used both between the owner and the architect and between the architect and his various consulting engineers. There is the "flat fee plus expenses" type wherein for a certain stated sum the architect engineer will produce a set of plans and specifications, supervise the work as called for, and perform all other duties as necessary to complete a satisfactory structure. This type of contract is recommended only for small projects. Before a project starts there is really no way of telling how much time and effort it will take and either the owner or the planner may become dissatisfied. A much more satisfactory contract is the one prescribing the fee (both from the owner to the architect and from the architect to his engineers) to be a percentage of the construction cost. In this case, if the owner requires a more expensive or ambitious building than was originally contemplated, there is no problem. However, when the fee is so determined, a word of caution should be extended to the architect engineer. The fee should be based on the original contract cost, and can rise if more work is done. However, if there are savings because the owner or planner or builder thought of omitting or economizing on a piece of work, the architect engineer should not be penalized by being paid a lower fee, because of his cleverness in working out such economies. There is another type of contract wherein the architect engineer is paid his payroll cost plus a multiple for overhead and profit. This type of contract is recommended for projects of indefinite duration or where the amount of work to be done cannot be determined in advance. In such cases, costs must be watched very carefully so as not to make the owner unhappy.

1B. SOME DUTIES OF THE ARCHITECT ENGINEER

1B1. *Field Inspection*

Depending on his agreement, the architect engineer has various duties to perform, some of which are essential. He must produce a set of drawings and specifications in sufficient detail so that a building contractor can build

CONTRACTS AND THE BIDDING PROCESS

a viable structure from them. Such drawings and specifications must comply with the existing zoning and building codes and the rules of any other agency that may have jurisdiction. It should be his duty to file such plans and to obtain the proper approvals and permits. If the plans deviate from the promulgated standards it is up to him to either obtain an exception or change the plans. He must keep the owner or client aware of what he is doing during the entire progress of the plan preparation. He should either defer to the owner's wishes for the use of certain layouts and material or he should explain why this should not be done for the good of the project. The periodic meeting between all interested parties during the plan preparation is again strongly recommended.

1B2. Shop Drawings

Another of the duties of the architect engineer that occurs during construction is the approval of "shop drawings." For instance, after the structural consultant designs the structure for strength, safety, and conformance

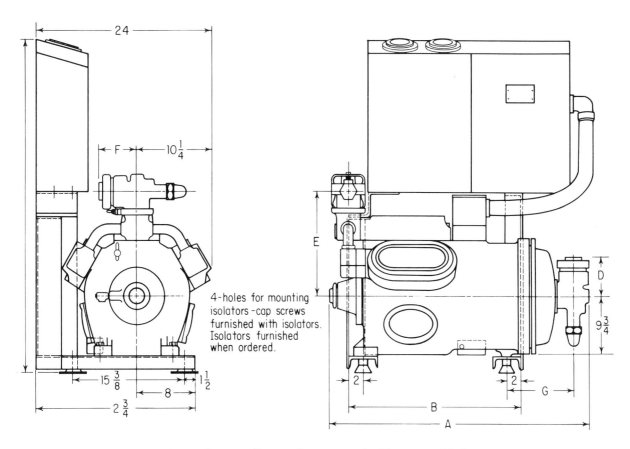

Example of a Drawing Showing Dimensions for Mounting a Machine.

with codes, the structural steel contractor takes these drawings and breaks them down to the actual dimensions of the pieces of steel, such as beams, columns, lintels, or angles which he will furnish, and shows how he will connect them in order to erect the structure. These drawings are called *shop drawings*. When the mechanical engineer designates a compressor or blower of certain capacity and performance, the manufacturer, whose equipment is purchased, sends in a drawing showing exactly how high and wide it is and how it is to be connected and where it must be supported and bolted down. Every trade must prepare such drawings and the architect engineer must study and approve them and coordinate them with his own general plans and with each other, so that when the material arrives on the job it will fit where it is supposed to go, and not cause delay and expense. He must also do this in time to meet the construction progress schedule and keep an accurate record of the progress of such drawings.

1B3. *Approval of Payments*

The approval of construction payments is also one of the important responsibilities of the architect engineer. When he approves a contractor's request for payment he is saying many things to the owner. He is guaranteeing that the work is in accordance with plans and specifications; that it is actually in place; and that the amount of the payment is such that the owner is, through the device of the retainer (which is usually 15%) reserving enough money to complete the job in the event of a contractor's default.

2. SELECTION OF A BIDDER'S LIST

2A. General Contractors

When the plans and specifications are complete and have been approved by the owner and all public bodies having jurisdiction, they are given to various contractors to bid. It is extremely important that prior to this the architect engineer and the owner have selected a list of contractors who are qualified to do this work. Choosing the right general contractor usually starts by the selection of a list of such contractors who are known for their integrity and ability to produce the quality of work that is desired. It may sound odd to state that in a list of contractors the quality of work can vary, and the question may well be asked, "Shouldn't every contractor be able to produce high quality?" Unfortunately the answer is "no." The contractor who builds tract houses is simply not geared to produce a fine residence; the contractor who builds speculative buildings for a quick profit is not fitted to build a monumental or high-quality institutional building. Neither his men nor his usual subcontractors' men are trained to spend the extra time that is necessary to produce the perfect fitting together of such a building. When several satisfactory contractors have been chosen, the next step is to send each of them a questionnaire. Such a questionnaire will ask each contractor for the depth of his organization; his financial stability; his key men and their backgrounds; the trades or portions of work he usually performs with his own men; a list of the projects he has built or is building, that are of the same general character as the one to be bid; the kinds of contracts he has usually entered into (see Section 3 in this chapter); the names of the subcontractors in the various trades that he uses for his jobs. The answers are checked by calls or letters to owners and other architect engineers for whom the contractor has previously worked, and not until then is he actually invited to bid. Usually, out of a list of seven or eight, from three to five are invited to bid. It should be mentioned here that while this procedure is a very important safeguard, it is not, and cannot be, perfect. In spite of such precautions a nationally known contractor can "fall on his face" by not having the proper local supervision, or an excellent local contractor can lose one or more key men in the middle of the job. The architect engineer must be prepared for such contingencies and must be able to overcome them by extra supervision and by carefully "minding the owner's business."

2B. Subcontractors

It was mentioned previously that one of the questions that a general contractor must answer concerns the subcontractors from whom he usually solicits bids. Since any of the bidders on the list can perform, the architect engineer and owner must be sure that they will accept the lowest bid before a bid is solicited. This must also hold for the subcontractors and the way to do this is to make sure that no general contractor solicits a bid from any subcontractor who is unacceptable. The major trades in which the subs must be so qualified are foundations, concrete work, structural steel, heating, ventilating and air conditioning, plumbing, electrical, exterior walls, elevators, and escalators. The architect engineer should know who the reputable men are, or should have ways of checking them.

The qualifying procedure mentioned above often cannot be applied to public work. Any contractor who can obtain a bond from a duly qualified bonding company is allowed to bid on city, state, or federal public work. This bond guarantees completion of the project in accordance with the plans and specifications. It is hopefully assumed that the bonding company has thoroughly investigated the qualifications and financial stability of the bidders before "putting their money on the line." Subject only to this, the public officials who solicit the bids must accept all comers.

CONTRACTS AND THE BIDDING PROCESS

3. CONSTRUCTION CONTRACTS

3A. How a Bid is Solicited and Assembled

Before going into the details regarding the types of contracts that may be made, it may be of interest to describe how the bid is solicited and how it is assembled by the general contractor. The letter of invitation that goes to the prospective bidder should give the following information: an actual copy of the form of contract he is expected to sign; the form in which he is to submit his bid; a schedule of alternates and unit prices he is to submit with his bid; a list of the approved subcontractors from whom he may solicit bids and usually six sets of plans and specifications. The letter of invitation will also state that the owner or architect engineer has the sole right to reject any or all bids or to accept any bid which in his sole judgment will serve the owner's best interests. The letter of invitation will also give the day and hour when the bid must be received.

When the general contractor receives this set of documents, he sorts the plans and specifications by trades and invites his subcontractors to bid on the plumbing, air conditioning, steel, etc. The information given to the subcontractor is based on the information given to the general contractor and the contractual terms should be understood in advance. There are printed forms of subcontracts which are often used, and the bidders know this. At a certain time before his bid is due the general contractor assembles his low subcontract bids and adds to them his own work cost and overhead and (hopefully) his profit. He then usually stays up all night going over the bids, the contract documents, and the plans and specifications to be sure nothing has been overlooked, and at the last possible moment delivers his sealed bid to the architect engineer. The competitive bids are usually opened by the architect and owner in privacy (except in public work where every bidder is usually present). Weight is given not only to the dollars involved, but also to such things as the organization of the bidder and his past record of successful projects. Although all bidders have been investigated and are qualified to bid and are presumed equal, "some are more equal than others"— as the saying goes.

3B. The Lump Sum

The lump sum, or fixed-sum bid, and contract can only be arrived at in those cases where the architect engineer has completed his plans to the last detail. The contractor must assure the owner that for a certain fixed sum, he will obtain a complete structure that is ready for business. This form of bid and contract is most often used in smaller jobs, and in private residence work, and must always be used in public work. As a matter of fact in some public work in certain cities and states the law calls for multiple bidding which means that various trades such as electrical, heating, ventilating, and plumbing are bid for and awarded separately by the public authority concerned. This is most definitely not recommended for private work or for that matter, for public work either.

Even in such lump sum contracts, however, there are always provisions for extra work and a list of unit prices for various changes the owner or architect engineer may wish to make. In fixed-price jobs, the contractor and his subcontractors are ever vigilant, to be sure that everything they do is exactly as called for by the plans and specifications. Claims for extras can become onerous.

3C. The Cost Plus Fixed or Percentage Fee

As directly opposed to the fixed-price or lump sum contract is the cost plus fixed fee, or cost plus a percentage contract. Such a contract must be entered into very carefully or "thoughtfully and advisedly" as the marriage ceremony says. As there is no competitive bidding, the team of owner architect engineer and general contractor must be in complete harmony. The contractor is constantly consulted during the progress of the planning and is called upon to give budget figures for the various trades involved and to revise them as the plans reach completion. Based on such final budget figures which *he does not guarantee,* the contractor is given a contract to proceed. Every dollar of payroll, or other expense, every subcontract that he enters into, and every extra order that he awards are subject to scrutiny and audit, and prior approval from the architect, engineer, and owner is always necessary for out of the ordinary expenditures. Nevertheless this type of contract calls for great care and is not advised unless the project cannot be bid competitively. There are too many known projects where the costs have enormously exceeded the budget.

3D. The Negotiated Contract

Sometimes when an owner and his architect engineer desire to retain the services of a certain general contractor, either because he has completed several successful projects for them, or he is uniquely qualified to build this particular one, they will negotiate a contract with him. Such a decision may be made at any time during the planning stage but as soon as it is made the chosen contractor must be called in immediately. He must attend all planning meetings and be made aware of exactly what kind of job is expected of him. He must be given enough time to carefully evaluate the completed plans and specifications, and to obtain bids from subcontractors. When he has done all this he names a sum for which he will complete the project. As the name "negotiated contract" states, this sum is subject to negotiation and there can be much "give and take." For instance, the contractor may say "If you eliminate so and so or use Smith's hardware or motors or elevators instead of Jones's as specified, I will reduce my price by so much." Or, "Do you really need two full basements? If you think it over and don't need them, you can save a large sum because this is a very poor foundation condition." Even after its closing, however, a negotiated contract can be changed by mutual consent to a guaranteed maximum type of contract as set forth in Section 3E below.

3E. THE GUARANTEED MAXIMUM PLUS FIXED FEE

There is a type of competitive bidding and a form of contract that combines the good features of the lump sum competition and the cost plus fee contract. This may be called the guaranteed maximum or guaranteed upset cost plus fixed fee bid and contract. In this case the list of selected general contractors bids competitively but the sums named are in several parts. First, the contractor names his fee; second, he states his guaranteed price; third, he states what participation he wants in savings from the maximum guaranteed price. In addition to the above he bids competitively on unit prices, which include everything that may be used during construction, but the quantity of which is unknown at this time. (The appendix gives a typical list of such unit prices.)

The contract is awarded on a combination of these figures, plus the contractor's organizational set-up and his guaranteed schedule of completion. The carefully drawn contract in the guaranteed maximum case must mention several key points. First, as in all construction contracts, it defines the work to be done. Second, it contains the list of the plans and specifications by title and page number. Third, it specifies the general conditions of the contract. Fourth, it states the fee to the contractor and his guaranteed maximum price. In this section the contract also states that all savings made in labor or material or in the final purchasing of the subcontracts *shall accrue to the owner,* except that the general contractor shall participate in the savings to the extent named in his bid (as mentioned above in the bidding procedure). Fifth, it describes the services the general contractor will render for his fee. Generally, this includes the knowledge and services of his entire organization. Sixth, it states in almost exhaustive detail the items that will be included in the cost of the work—to name a few: the wages and salaries of named employees (by category); the general contractor's personnel and the material used by them; the cost of duly authorized subcontracts; the cost of renting equipment, traveling expenses, telephones, telegrams, site office rental, sales taxes and use taxes, fees, insurance, and permits. Seventh, it prohibits the contractor from entering into any subcontract without the prior approval of the architect engineer and the owner. In this case, the general contractor must exhibit his subcontractors' bids and they must be from previously qualified bidders. Eighth, the contract explains how changes in the work are to be submitted and accepted and mentions that as the cost of the work goes up the contractor's fee is increased by the percentage named in his bid. The contract then goes on to mention the manner in which payments are to be made. There are various legal safeguards which protect the owner against default on the part of the general or subcontractor. This form of bid and contract seems to offer maximum safety for the owner and rewards the general contractor for good work by giving him participation in the savings.

3F. THE SCOPE CONTRACT

The scope contract is a term used for what is really a guaranteed maximum contract, but which is based on incomplete drawings and specifications. Quite often for large projects, whether they are factory, multiple dwelling, or office building, where the land cost is high, or when an important tenancy is involved, or a manufactured product must be quickly put on the market, the owner and architect engineer, after choosing certain highly qualified general contractors and subcontractors, will ask for competitive bidding on a guaranteed maximum plus fixed fee basis on incomplete or outline plans and specifications. The plans may not be fully developed and detailed for months after the construction is started, but the contractor and his subs dealing with the expert architect engineer know what they are expected to produce and such jobs can go very smoothly. In such cases after he has entered into a guaranteed maximum contract the general contractor and his subcontractors, having guaranteed the cost from incomplete drawings, may "look over the architect engineer's shoulder" as he develops the plans (and object if he is overdeveloping). This scope guaranteed maximum contract is only as good as the parties that sign it, but it can save many months of waiting for plans and specifications to be completed.

3G. THE GENERAL CONDITIONS OF A CONTRACT

The general conditions document forms a part of every construction contract. It is devoted to a detailed listing and explanation of the duties and rights of all the parties concerned in a construction project. It can be qualified as conditions warrant but it never varies in its basic definitions of the interrelationship of the plans and the specifications; of the owner's duties to furnish certain information such as surveys and the owner's right to terminate a contract; of the architect engineer's rights to inspect the work, to issue binding

CONTRACTS AND THE BIDDING PROCESS

instructions to the contractors, and of his duties to pass upon shop drawings and the quality and progress of the work; of the contractor's rights to claim compensation for extra work, to terminate a contract, and his duties to coordinate the work of his subcontractors, to protect work in place, to furnish certain insurance protection to the owner, to take certain safety precautions, to provide supervision. In addition, many specially tailored general conditions documents describe in detail such things as temporary lighting and heating during construction, safety precautions, barricades and sidewalk protection, fire protection and watchman service, temporary enclosures, cutting and patching, sanitary facilities, and temporary elevator service. The complete contract documents include the general conditions as well as the plans and specifications and the contract itself.

3H. Use of a Cost Estimator

The use of a cost estimator is a univeral practice in large projects and frequently in smaller ones. The cost of the construction is one of the first things the owner and architect engineer must discuss. There is always a budget. The owner tells the planner what he expects to pay for the completed structure. Sometimes the planner is told that the budget cannot be exceeded. Then the architect engineer would be wise to employ a qualified estimating firm to constantly check his plans as they progress and to immediately advise him if he is overdesigning or using methods and materials that may exceed the budget. Frequently cost estimators are employed by the owner. It is possible, especially if there is a close relationship between the owner or the architect engineer and one or more general contractors, to employ such contractors as cost estimators. Such firms generally have their finger firmly on the pulse of the market and if they are also to be placed on the bidders' list, the advantages of such an arrangement are obvious. In any case, even if a general contractor is used he is, of course, paid for his services so that there is no obligation on either side. There are a suprising number of instances in which the estimating general contractor, who is presumed to have the "inside track" does not get the job.

CONTENTS

1. EXCAVATION
    A. Test Pits and Test Borings
        1. The test boring
        2. Underlying soils that may be encountered
    B. General Excavation
        1. Soft soils
        2. Rock
        3. Dewatering
        4. Soil stabilization
    C. Underpinning and Sheet Piling
    D. Pier Footings and Pits
    E. Problems that May be Encountered

2. FOUNDATIONS
    A. The Design, Methods, and Materials Used for Footings
        1. Spread Footings
        2. Caissons
        3. Piling
    B. Waterproofing

3. INSPECTION OF ADJOINING PROPERTY BEFORE WORK STARTS
    A. Use of the Seismograph

5

EXCAVATION AND FOUNDATIONS

In Chapter 3 (Section 2A2) the author has said: "The architect also knows that his structural designer will be able to design a suitable foundation for almost any structure within reason." This is a broad statement but it is true. Of course practically no one would want to build a 100-story building on a bottomless swamp but in lower Manhattan in New York City they have almost done that very thing.

The architect engineer, in the normal course of events, unless he is in a large city with a firm that designs large structures, will rarely run into such difficult foundation problems. Large or small, however, the duty of a foundation is to support a structure and before a foundation can be placed, the underlying ground must be excavated to accommodate it. We shall therefore start with a simple structure and foundation and go on from there.

1. EXCAVATION

1A. Test Pits and Test Borings

Before designing the foundation for any structure some investigation should be made of what lies beneath the surface. Even for a single-family residence this should be done especially if the owner wants a full basement. It costs very little to dig a test pit to determine the ground water level and the

EXCAVATION AND FOUNDATIONS

underlying soil. There are instances where private homes have been built on the site of an old swamp and have tilted and cracked. A test pit would probably have shown peat or black mud beneath the surface fill. It is also advisable if the terrain is of that character, to check for ledge rock. It can come as an unpleasant surprise to find a dug cellar pit full of water or a portion of it not useable because of solid rock which has to be blasted. It could be easier and much less expensive to move the house location or to arrange with the contractor to waterproof or raise the basement level, before construction is started. Moving the location is not possible, of course, except where there is a great amount of land, and property lines are not close to the structure. In all other cases the test boring to determine the character and bearing capacity of the underlying soil is an absolute necessity. In a small business or apartment structure, if there is intimate knowledge of what soil and water conditions were encountered in adjacent buildings, there may be enough information available to the designer and the contractor who is to bid on the job. But even here they must accept this information on faith because rock strata and clay beds have very peculiar ways of tilting upward or downward within a very short space.

Every building code publishes a list of approved bearing capacities of various materials. These range from bedrock such as granite, trap rock, or hard limestone that can carry a load of 100 tons per square foot to such materials as schist and slate in sound condition at 40 tons per square foot, to fine sand (confined) at two tons per square foot, and soft clay at 1.5 tons per square foot.

1A1. *The Test Boring*

Although the making of test borings is a comparatively simple procedure it should be done by people who know what they are about. A boring rig usually consists of a structure holding a hollow steel pipe within which is another hollow pipe. This pipe is mounted vertically in guides, has a hardened toothed bit at its bottom and is driven downward into the soil. As it is driven downward, it forms a circular hole into the earth and a core forms in the middle of the hollow pipe. When this drill hits rock it is rotated to form a rock core. This core can be retrieved at certain intervals to show what kind of soil or rock is being encountered. The particular weight (140 pounds) and the height from which it falls (30 inches) in order to drive the pipe into the ground is in accordance with a standard set by the American Society for Testing Materials (ASTM). When the pipe reaches the maximum depth of the excavation, the amount of give of the pipe, as it responds to the hammer blows, is an indication of the bearing capacity of the soil, which can then be ascertained from prepared tables. With this information at hand and with the core itself available for inspection, the structural designer can safely design footings and foundation walls of the proper size to support the structure and to hold back external pressure. He can also usually tell from an inspection of core whether he is likely to encounter water, so that he can also take precautions against this condition.

1A2. *Underlying Soils that May be Encountered*

The subsoil conditions that are encountered vary widely over the country and even within blocks of each other in a city. On Manhattan Island in New York City, there are locations where bedrock is at street level, and, within a thousand feet, has sloped down to 40, 50, or more feet below this level. There are other locations here, especially in the southern portion of Manhattan where there is almost no bottom at all. In the borough of Brooklyn in New York City, there are certain places where there is terminal moraine (pushed there by the great glacier), which consists of compact sand and loose boulders and is usually two and three hundred feet deep so that any foundation, except on sand, is impractical.

In Chicago, over a great part of the city, there was originally a great marsh, which eventually dried out and left about 80 feet or more of silty, clay soil over bedrock. In Los Angeles, an ancient sea left clay and silt stone, which in the 10 million or so years since its deposit, has been covered with an overburden of about 20 feet of silt and clay soil. After such overburden is removed, all foundations here must be built on this siltstone material. There are cities where there are inclined planes of moist clay which can slip upon each other if any great burden is placed upon them. All of these conditions must be thoroughly studied and understood by the designer before he puts pencil to paper. In a recent construction project that the author encountered, there was a question not only of the bearing capacity of the soil but also of an underground stream which, according to old maps, might be flowing right through the site. In this case, a pattern of test borings was made over the entire site and some of these test borings were left open with a perforated steel casing in place to determine the amount of water that flowed into such a test hole. The boring cores and the pattern and rate of water flow were studied by a consulting geologist who worked with the structural designer. The general contractor, who was awarded the job, knew what to expect and made the necessary preparations for it. These studies took time and money but the excavation was performed in record time with no trouble. The project received such an impetus from these studies, that it has been a highly successful one with a saving in money and time of many times the original extra expenditure. This case shows that careful preparation and intimate foreknowledge is very important for the planner's reputation as well as the owner's satisfaction.

EXCAVATION AND FOUNDATIONS

1B. GENERAL EXCAVATION

Before beginning a discussion of the materials and methods used in general excavation it would be well here to quote from a typical specification for this work. (1) "Excavate down to levels as required for footings, walls and piers, and mechanical and electrical installations. (2) Provide sheet piling or underpinning to prevent displacement of earth and of adjacent structures, and to protect workers as is found necessary. (3) Provide pumping to maintain subgrades in dry and firm conditions." These three sentences combined with the necessary drawings and test boring information, tell the contractor what his duties are.

1B1. *Soft Soils*

The excavation of soft soil is quite simple if it is just clay, sand and gravel, or a combination of these. For the small structure this can be done by a bulldozer or shovel dozer that scoops out the soil and pushes it into a pile (don't forget to save the top soil) or dumps it into a truck. For an excavation that goes down several levels a power shovel or a bucket crane is used. In some sandy soils one is likely to encounter large boulders which must be blasted, but generally earth excavation presents no problems. If the soil is stable enough and the site large enough so that the perimeter banks of the excavation can be left in a natural slope which will stay (this is the angle of repose or stabilization which varies for different soils), then the owner and contractor are very fortunate. Generally, however, the perimeter banks have to be supported by sheet piling.

1B2. *Rock*

The removal of rock from a large excavation presents few difficulties. The technique now being used makes this almost an assembly line process. In the case of rocks such as soft and broken bedrock, soft sandstone, or so-called cemented conglomerates, a great deal of the operation can be performed by bucket crane or power shovel, but when hard rock requiring blasting is encountered, the process, while still comparatively simple, requires more expert attention. The usual procedure is to drill a pattern of holes, the depth of which, and the distance from center to center, being determined by the hardness and structure of the rock. Each hole is then loaded with dynamite and the blast is set off electrically. In order to prevent a serious tremor when a number of holes are blown at once, the blasting caps are fitted with a device delaying the action 1/1000 second between sticks and this creates a series of explosions instead of a single massive one.

If the rock has been blasted expertly, it smashes the mass into small chunks that can be lifted by a power shovel into trucks for disposal. The expertise with which such blasting is done can be noted in any large rock excavation for building or highway, by observing the series of half round drill scorings where the rock has split away from the surrounding mass as neatly as if it were cut with a knife. This process is known as line drilling and is done this carefully only at the property or building line. Usually in cities, where there are surrounding structures close by, it would be wise to employ a "retained specialist" firm. They inspect buildings for damage, advise on the blasting, and can furnish one or more seismographs placed at strategic points around the perimeter of the excavation and interpret the seismographic readings.

1B3. *Dewatering*

An important examination that must be made of subsoil conditions is to determine whether water will be encountered during the excavation. There are possible underground streams everywhere, and often in the most unlikely places. This is apt to be so in established areas where the streets are paved and streams have long since been covered with earth and fill. But even in the country there are underground springs which are not apparent to the eye and which can cause havoc with the owner's requirements for a basement, or even with the digging for a foundation. There are some cases where, instead of underground springs, the water table is higher than the depth of the completed excavation.

This water obviously must be taken care of. A spring or a small underground stream can be diverted by the use of piping laid in a trench filled with crushed stone. This will divert the water before it gets to the excavation. In a large excavation, an underground stream can be led through the excavation by means of piping but generally this is quite difficult to do because old stream beds are not very well defined and one cannot be sure that expensive trenching and piping will do the job. In such cases the best thing to do is to keep the excavation dry while general excavating, trench and pier hole digging are going on and then build the basement slab and foundation walls with sufficient strength and water resistance to keep the water out of the building. It should be remembered here that water will take the path of least resistance and if it can be led away from the building basement, it will be pleased to not have to force its way through a concrete floor slab or foundation wall.

There are several methods of keeping an excavation dry, such as lift pumps which are used to pump water out of collection pits after heavy rains or where a small amount of water is present, but when the bottom of the excavation is below the water level because of a stream or a high water table, the use of well points is recommended.

The wellpoint works on the physical principle that when a vacuum is applied to the top of a column of water, atmospheric pressure will push the water up the column. What happens is that a series of pipes which are pointed and perforated at the bottom are inserted into the earth at a level

EXCAVATION AND FOUNDATIONS

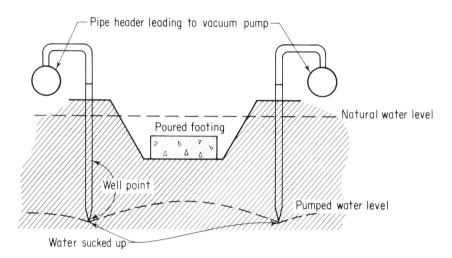

Well points lower the water level until footings are poured and set

A Wellpoint Installation.

below the bottom of the finished grade of the excavation. A vacuum pump is connected to these pipes through a header pipe as shown and these pipes simply suck the water out of the underlying soil and keep sucking this water out until the excavation has been completed and the footings, wall footings, or basement floor slab has been poured. The amount of wellpoints to be used is determined by the quantity of water estimated by subsoil examination, but many contractors will protect themselves by stating that they will supply only so much pumping for a defined area for so many days. After that they want to be paid extra. The original estimate of the water that may be encountered is therefore important.

1B4. *Soil Stabilization*

Some soils that are encountered during excavation are so unstable that it is very difficult to keep them in bounds so that excavation can continue. They may be water impregnated, or slippery clay or sand, or a combination of these and gravels, etc. Such soils can be found anywhere and it is well for the architect engineer to have a general idea of the procedure to follow. There are experts in this field who will be pleased to advise, especially if their material is used (sales engineers). The unstable soil must be solidified and this is most often done by the use of chemical soil stablizers. In general, wellpoint pipe is driven into the soil to a level below the finished bottom grade and a predetermined amount of a solution of sodium silicate is pumped down and this is followed by a solution of calcium chloride. The two chemicals combine with the soil to form a sort of silicate sandstone which is hard enough to hold itself up, and will not slide. There are other chemical soil stabilizers. The process is expensive, however, and should be used only as a last resort.

EXCAVATION AND FOUNDATIONS

1C. Underpinning and Sheet Piling

In excavations where the soil that is removed consists of such materials as gravel, sand, soft earths, soft clay, very crumbly rock, or other materials of this consistency, it is necessary to perform one or two operations to assure the safety of workmen, of surrounding buildings and other property, and to prevent serious damage within the excavation.

The first of these operations is called *underpinning*. In most communities, when the foundation plan calls for the basement of a building to go below the level of the footing of surrounding buildings, it is the duty of the owner and his architect engineer to design and build a substructure to hold up the foundations of such buildings. If information as to subsoil conditions and the state of the foundations of the adjoining buildings are not readily available, then such work is done by the contractor on some sort of "cost plus" basis.

Such underpinnings can be a fairly simple operation consisting of an excavation under an adjoining building foundation at intervals, and the pouring of concrete under such footing until the entire wall is underpinned

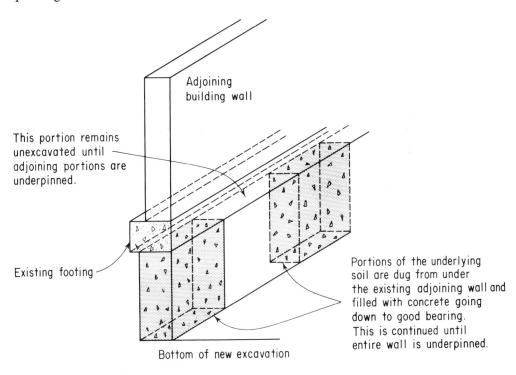

This Is An Illustration Only. Such Work Requires
Expert Knowledge Regarding The Condition Of The
Wall, Of The Soil, Of The Loading And Other Matters.

Typical Underpinning Operation.

EXCAVATION AND FOUNDATIONS

with the concrete going down to the depth of the new excavation. Generally this can be done when the bearing soil is capable of supporting this load. In cases where it is not, the underpinning has to be done by driving piles to support such adjoining walls by means of brackets, or by jacking piles under the walls, or by spreading the footing under these walls. In any case, however, they must be thoroughly supported. Sometimes it is necessary to underpin public streets or other underground facilities where the excavation is deep, the soil is crumbly, and the building line is adjacent to them. This is fairly rare because the architect engineer usually takes precautions against such street collapses by sheet piling or similar supports.

Sheet piling can be simply described as the support furnished to the surrounding earth to keep it from collapsing into the excavation, whether it be the general excavation, footing trenches, pier holes, or pits.

The material of which such sheet piling is made and the method by which it is installed varies considerably depending on the depth of the excavation, the character of the soil, and such things as the height of the surrounding buildings, even though they are not immediately adjacent to the site; the amount and kind of traffic in the surrounding streets, and such considerations as the possibility of earth tremors or possible torrential rainfalls in the area. The architect engineer who has the borings and knows approximately the underground water situation, then designs the most functional

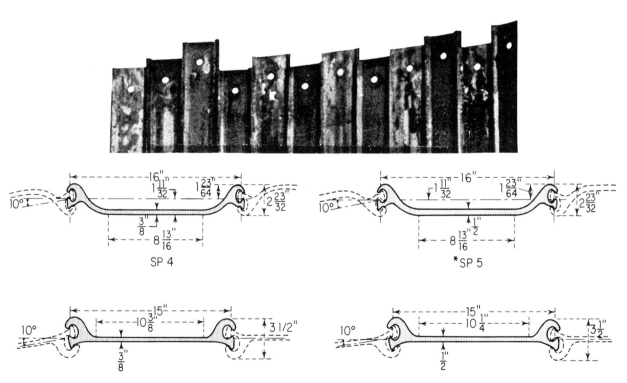

Steel Sheet Piling.

sheet piling support. For excavations where the surrounding soil is not very stable, steel sheet piling is very often used. Such piling is usually driven down immediately against the earth bank as the excavation proceeds, and the piling is always seated down below the level of the excavation. Steel sheet piling usually consists of interlocking sections which form a practically waterproof bounding wall and are supported by braces from inside the excavation. A typical installation is shown in the sketch.

In some cases where the soil is of reasonably consistent clay, or soft bedrock, or hardpan, a bank may be formed at the proper angle of repose as approved by the soil engineers. Sheet piling need not start until the bank has reached the edge of the actual excavation. In this case, timber sheet piling consisting of long planks driven down and held by cross-bracing and bracing from the excavation can be used instead of steel sheet piling.

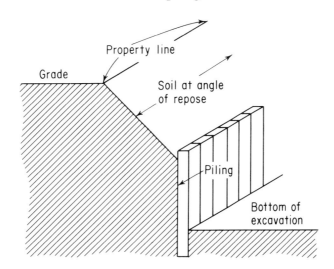

Piling at Bottom of Slope.

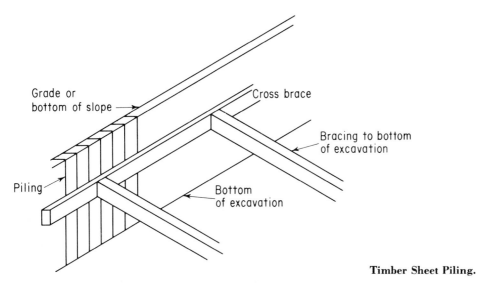

Timber Sheet Piling.

EXCAVATION AND FOUNDATIONS

There are many interesting variations of piling such as the one shown, where horizontal timbers are held by the flange of an H-column called a *soldier pile*, which in turn is held up by cross-bracing or lagging. These cross-bracing timbers or steel H-sections are shored by bracing from within the excavations.

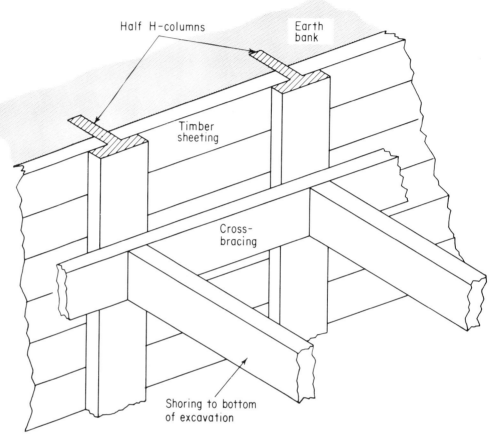

Timber and Steel Combination for Holding an Earth Bank.

The author knows of a construction job where a reasonably good hard clay bank was held by soldier piles. These are steel or timber vertical members held against the bank at suitable intervals by tie back-bracing. In this case, a hole is drilled at a downward angle into the bank behind the soldier pile. A steel rod with a bell-shaped attachment at its end is inserted into this hole. Concrete is then forced into the hole surrounding the rod, and the bell-shaped attachment at its end serves as a plug which is held down by the concrete poured from it to the edge of the bank. The soldier beam is then bolted to the end of this rod.

This method of bracing has the great advantage of not cluttering up the excavation and can be used in cases where the earth is reasonably firm and has some capacity of resistance to the possibility of the concrete plug being pulled out by the pressure of the earth against the pile. For the protection of pits, or trenches, or pier holes which are not too far below the level of the general excavation, timber sheeting and bracing is used most often.

The importance of the method by which the excavation is to be protected and the exact knowledge of the bearing capacity and stability of the surrounding earth during excavations cannot be overemphasized. The author knows of several cases where the streets or buildings have collapsed into an excavation. We have all heard of workers being buried alive under collapsing earth. Such tragedies can be avoided by proper design and supervision.

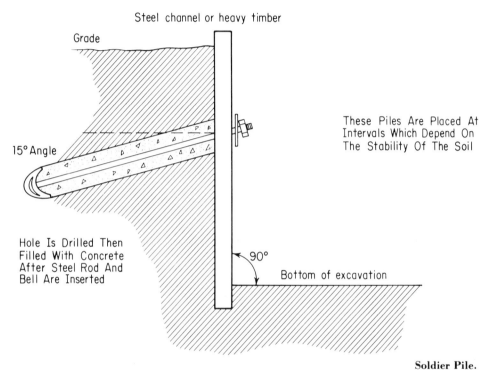

Soldier Pile.

1D. PIER FOOTINGS AND PITS

When the general excavation reaches the finish bottom grade, excavation for pier holes, pits, and foundation wall footings commences. For the two- or three-story private residence, apartment house, and business building with basements, such footings should be dug to solid undisturbed inorganic soil which, of course, is known to be able to carry the load to be placed on it. Where there is no basement, such footing excavation must go down at least below the local frost line and certainly to undisturbed soil.

In the case of hard rock excavation the wall or pier footing must go down far enough below the general excavation to prevent any side slippage. If blasting is required, the rock sometimes breaks in a sloped pattern. Then the bottom of the hole must be "stepped" to provide horizontal bearing for

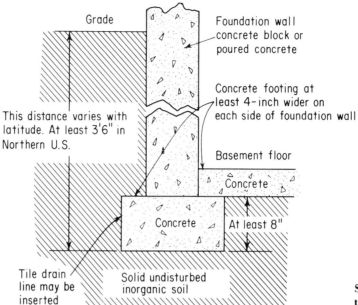

Small Building (up to 3 Stories) Foundation Footing.

the footing. Very often the designer will specify drilling whereby drill holes are made at the bottom of the hole and heavy reinforcing rods, which will extend into the footing, are cemented into the rock.

The owner and architect engineer who find rock or other good bearing near the bottom finish grade are fortunate. This is not often the case however, and several methods are used to get down to solid bearing for the piers which support the columns. It should be noted here that while in low, wall bearing buildings[1] (up to five stories), the foundation walls carry the weight of at least the outer walls; in higher buildings, the foundation walls are simply the perimeter walls of the portion of the structure which is below grade, and it is the piers that hold the weight of the building by supporting the reinforced concrete or steel columns, which in turn support the structure. Therefore, it is the footings for the piers that must go down to solid bearing where the upper layers of soil cannot hold the weight to be superimposed on them, or where it is more economical to go down to rock or hard pan than to build huge spread footings.

Several methods for getting down to such solid bearing for dry holes are currently in use. The earth auger is exactly what its name implies. It is a large auger blade that looks exactly like an enormous wood drill bit and it usually is set on a rig that revolves it and presses it into the earth. The soil that is brought up (like wood shavings) is cleared away by hand or a bulldozer blade. Such drilled holes can go down as far as 80 feet when the soil is

[1] In a wall bearing building, the perimeter wall, usually masonry, carries part of the building load.

self supporting and the diameter of the hole is not much more than three feet. In such cases men are sent down to the bottom to enlarge the hole at the bearing point before the pier is poured. Where large holes are necessary, an auger of large diameter can be used to start the hole but then the sides of the hole must be shored as it goes down. Usually this is done by means of timbers that are cross-braced. One ingenious contractor in Chicago where rock was about 80 feet down, purchased old railroad tank cars, cut both ends off, and used them as casings for the pier holes. As the excavation by hand and bucket crane proceeds, the casing goes with it. Such casings are often pulled up when the excavation is complete and the hole is being filled with concrete. Sometimes because of the underground water condition, good bearing can be reached only by means of compressed air caissons. The use of such caissons is avoided wherever possible because of the expense and because men have to work under heavy air pressure. The author cannot recall any recent excavating work on land where such caissons have been used. Usually if good bearing cannot be reached by open excavation other methods are used. This will be explained under "Foundations" in this chapter.

1E. PROBLEMS THAT MAY BE ENCOUNTERED

In numerous instances, excavation and foundation work presents unique problems and the architect engineer has to be ingenious to solve them. In the late 1960s, as land has become scarce especially in central city areas, entrepreneurs have purchased air rights over railroad yards, highways, and even railroad stations. For a negotiated annual rental set by a long-term lease, these entrepreneurs have the right to construct a building over such property. The architect engineer team has to plan the structure so that its foundations may be placed with the least inconvenience to traffic. Column spacing must be such that they fit between tracks or through platforms or at the center isle of a highway. Foundation holes for column footings must be dug while the railroad or highway is in constant use. Such work has been done successfully over both the Grand Central and Pennsylvania railroad yards in New York City, and over several railroad yards in Chicago, and undoubtedly in other cities.

In a recent project, no one knew that the huge (and very hard) concrete footings of old railroad trestles were buried under several feet of soil and could be broken apart and removed only by heavy blasting. Such blasting was impossible because of a nearby building containing very expensive and delicate machinery. It was a pile job. (Piling was to be driven down to rock to support the column footings.) The contractor and architect engineer came up with a solution whereby a heavy drill rig drilled large diameter holes through the old concrete piers and the piles were driven through these holes and the soil beneath down to the underlying rock. This method could be also used in the case of very large boulders.

EXCAVATION AND FOUNDATIONS

In Chicago, where rock is at least 80 feet below the remains of a great marsh which has hardened into a clay silty soil, one sometimes runs into underground pockets of marsh gas or methane which has formed from rotting vegetation and become trapped by the surrounding soil. Strict safety rules are enforced where such pockets may be encountered because methane is explosive when mixed with air. No smoking is allowed near the excavation and the diggers must use explosion-proof lights. They use sparkproof metal equipment in case they hit a piece of rock, and the men have a two-way telephone connection to the surface. They also carry with them a chemical that turns color when methane is present and arrangements are made so that when this occurs, the men can be hauled up immediately.

There may be instances where although the tested and approved bearing capacity of the soil under the foundation is known, there is doubt about the stability of the soil in the entire area with respect to its possible settlement *after* a structure has been erected on it. In such a case "swell and settlement markers" may be used. Such markers, which are firmly embedded in underlying soil that is to remain undisturbed throughout the construction, are used to accurately (within .0001 foot) measure the "relaxation" or swell of the underlying soil as its overburden is removed, and its settlement as the weight of a structure is placed upon it. The expert geologist can draw a curve from the readings to foretell when final stabilization will take place. The expense of such an investigation cannot begin to approach the cost of repairing the damage that could be caused by severe settlement cracking.

2. FOUNDATIONS

The purpose of the foundation is to transfer the weight of the structure by various means to the underlying soil so as to produce a stable situation in which the soil is capable of bearing the weight placed upon it without settlement or slippage.

2A. The Design, Methods, and Materials Used for Footings

2A1. *Spread Footings*

In order to perform its duty the foundation must be designed so that it distributes the weight of the structure evenly to all parts of the underlying soil. In cases where the underlying soil is capable of bearing the load to be placed on it, and such soil is reasonably close to the level of the finish basement grade, the footing for the supporting columns is placed directly into

EXCAVATION AND FOUNDATIONS

the excavated pier hole. Such footings consist of controlled concrete which is usually designed to bear 3000 to 5000 pounds per square inch (psi) in compression. They are reinforced with deformed steel bars in accordance with the engineer's design. Where hard rock is immediately available below the surface of the general excavation, such a footing need be only large enough to transfer equally the loading on each column to the rock below at the rate of 100 tons per square foot. Except in very heavy structures, such concrete column footings are comparatively small. In schist or slate which can bear 40 tons per square foot, the concrete footing must obviously be 2½ times as large as the one on rock, and it must be correspondingly deeper, and contain more reinforcing. The footing really gets large and deep when it is on clay or on confined sand. In such softer soils where approved bearing soil may be found at different levels, or where a footing is near a pit, the design must permit a stepping up or down of footings to prevent slippage. In such cases they may

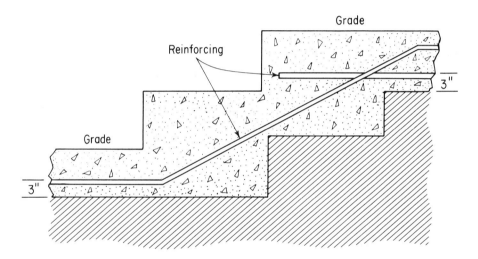

The Depth And Width Of The Steps Depend On The
Kind Of Soil And The Loading On The Footings

Stepped Footing.

become a single footing. Account must be also taken of such factors as wind pressure on the upper structure, earthquake exposure, and heavy vibration from machinery that may be used in the structure. To give one an idea of how massive such footings can be, an example of spread footing size and design for a tall building set on hard clay follows. For the main bearing columns these footings are made of 3000-psi concrete and are either 22' 0" × 22'0" × 6'6" deep or 17'0" × 24'0" × 4'0" deep. They are reinforced with rods in both directions at both top and bottom.

EXCAVATION AND FOUNDATIONS

2A2. *Caissons*

Where excellent bearing such as rock is reasonably near the surface and the structure is so large and heavy that it must bear on such material, the designer arranges to reach this rock by either caissons or piles. The excavation process by means of caissons has been previously described. When the excavation is completed to good rock, it is required practice to step the rock to a horizontal plane if it slopes, and to cut into it at least 12 inches so that the concrete footing sits securely in place. The placing of reinforcing rods is the next step. In a case where 8-foot diameter caissons went down to rock about 80 feet below grade and held a tall building, the design called for 34 heavy reinforcing rods placed vertically in a circular pattern from top to bottom and bound horizontally by heavy wire at 18-inch intervals. (This is an example and cannot be used as a guide. It merely shows the magnitude of the work.) The cylinder was then filled with 5000 psi concrete carefully placed and vibrated into place to provide a solid mass. In this case the shell of the caisson was left in place. The author knows of one case where the caisson shells were to be reused. This was done by pulling up the shell slowly after the concrete was poured but before it set. Unfortunately, in this case, the shell was pulled too fast and earth and clay pressed into the soft concrete and created large voids in the footing. The condition was not immediately discovered and it cost many hundreds of thousands of dollars to correct it. This does not mean that expensive caisson shells should not be reused. It does mean that extreme care should be exercised when this is done.

2A3. *Piling*

The use of piles to support wall and column footings is quite commonly specified in cases where it is difficult to reach good bearing soil by other means. The most common use of piles is for structures over water such as piers and bridges. The Chesapeake Bay Bridge which runs over many miles of open water is a classic example of the use of pile foundations. Hundreds of hollow prestressed concrete piles were driven into the bay bottom, filled at the top with concrete plugs, and capped with poured concrete to form a base for the bridge.

For structures on land, building codes are explicit as to the kind and strength of piles to be used, how they are to be driven, and what their allowable bearing capacity can be. Such codes will specifically describe a timber pile; the kind and quality of wood; the minimum diameters at various points in its length and the use of a preservative; or a concrete pile which must be of a certain compressive strength (usually not less than 3000 psi). The codes also describe steel pipe piles which are to be filled with concrete; rolled structural steel, or H-column piles, or composite piles which may be of concrete and

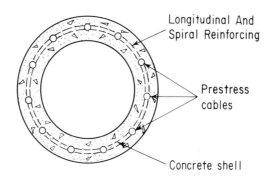

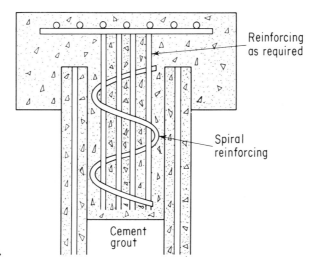

A Concrete Pile Cap

Raymond Hollow Concrete Pile.

steel, or concrete and wood. The architect engineer's specifications for the pile driving and the usual code requirements are very strict. In all cases, where the pile does not reach solid bottom, it is the friction of the soil on the pile that holds the entire structure and it must be certain that this friction will do the job. The specification may state that a pile driver must use a certain weight of hammer falling a certain distance to drive the pile until the last specified number of blows drive it only so far. The last few blows may drive it only a small fraction of an inch or to *refusal* (where it doesn't move at all). The specification (or code) goes on to state that certain piles may carry only certain maximum loads and must be tested to twice the allowable loading. For instance a typical code states that if a pile settles not more than .01 inch under a certain load in 24 hours then the allowable load may be *only half that*. Pile driving and pile testing is always witnessed by experts from the designer's office and the public authority.

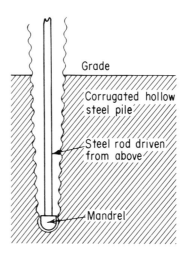

A Tapered Hollow Steel Pile.

There are many types of piles used and many methods of driving them. There are steel pipe piles of many shapes—straight, tapered, or stepped-driven down from the top or by a mandrel (a steel rod which bears on the bottom of the pile and transmits hammer blows to it from above.). All of these are filled with concrete after they are seated. There are, of course, timber and concrete piles which, as stated above, must be driven and tested to certain specifications. One type of pile used depends very little on skin friction and instead makes its own spread footing. The "Franki Pressure Injected Footing" drives a plug of dry concrete or gravel which is contained at the bottom of an open ended tube by means of a number of very hard blows. As the plug is driven down it draws the pipe with it until it reaches a soil which tests have shown can bear the weight to be placed on it. At this point the plug is driven down and out of the pipe and zero slump concrete[2] is rammed into the soil beneath the pipe until it forms its own spread footing. When the spread footing, which has been formed by the enormous force of the pounding, is of sufficient size to bear the weight to be placed on the footing, the cylinder above it is filled with concrete of the designed strength and is reinforced if necessary. This compressed footing does the work of the skin friction which supports the ordinary pile and can be used in unstable soils.

Of the many other types of piling, two will be mentioned here. These are used where it is impractical to use ordinary caissons to reach solid bearing. The first is the drilled-in caisson as developed by J. H. Thornley (Drilled in Caisson Co.) and Spencer, White and Prentis. In this case a steel shell is driven down to rock and firmly seated on it. The earth and muck are removed from the shell. A socket is then drilled or pounded into the rock for a distance of at least 10 feet. When all drilled material is removed, a very heavy H- or double HH-column (depending on load) is lowered into the hole and

[2]This is concrete that contains just enough water to make it harden but is so stiff that a test cylinder can stand by itself without support and with no slump.

EXCAVATION AND FOUNDATIONS

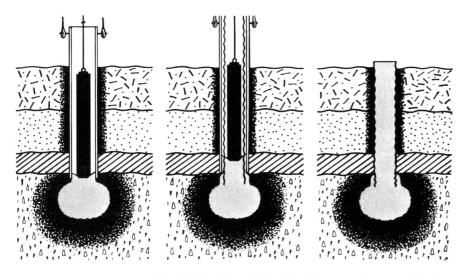

Franki Footing. Left: The Pressure Injected Footing is Formed. Center and Right: The Concrete Shaft is Formed from the Footing to Grade.

then the entire hole is filled with high-strength concrete. This type of pile foundation is very useful when rock is more than 100 feet below the surface and the pile has to carry heavy loads. Such a single pile (42 inches in diameter) can carry a load of 3700 tons. The second piling form to be mentioned here is the H-piling. This piling consists of a rolled structural steel H-column which is driven down to rock and is probably as widely used as any form of piling. The driving and bearing capacity of this H-piling is thoroughly controlled by the design as well as by building codes. As for all piling this one is driven in clusters of as many as is necessary to support the superimposed weight—a typical specification will allow 150 tons of weight per pile. In some city codes they do not allow less than a cluster of three to hold a foundation wall or a column footing.

In the use of any piling the design always calls for a pile cap (See page 70) which is constructed of reinforced concrete and which takes the place of the footing on soil as previously described. The tops of the wood or steel pipe or H-column piles are cut evenly at a specified distance below basement grade level and above footing level, and a concrete "block" is poured over and around them. The block is of such dimensions as to completely encase them and to distribute the weight from above evenly to each pile.

In conclusion, the proper use of any piling depends on a close study by the architect engineer of bearing capacities and stability of soils, the distance from the finish grade of rock or other good bearing material, and even the chemical composition of underlying soils. It all goes back to proper investigation and informed study of what lies below his proposed structure.

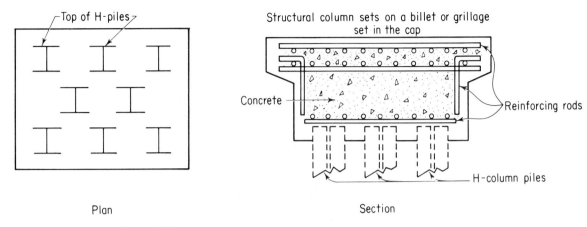

Typical Pile Cap.

2B. WATERPROOFING

The proper waterproofing of the foundation wall and floor which lies below grade and where investigation shows the presence of water, requires that the architect engineer have some knowledge of the water pressure to be encountered. As stated previously under "Excavation" (1B3), water will not try to force its way through concrete floors or walls if it can be led away from them. This is not always possible especially where there is a high water table, or the structure is close to tidal waters or rivers whose level may change with the seasons.

First the basement floor slab should be reinforced and of sufficient thickness to resist the hydrostatic pressure. The concrete of the floor slab should be waterproofed with an integral waterproofing compound, and should be of a dense mixture, and thoroughly vibrated or puddled into place. It should also be placed over a bed of crushed stone or gravel which may be as much as 12 inches deep and which is covered with a thin layer of waterproof material before the slab is poured over it. This crushed stone serves as a cushion against water pressure and the water can flow under the floor and around the outsides of foundation walls and into sump pits. In some situations, however, the water simply has no place to go and there are cases where even very heavy floor slabs must be pierced by pressure relieving pipes which project above the floor for several feet. Whenever the ground water level rises to the height of the pipes above the floor level, the pipes spill the water onto the floor. This happens very rarely and is inconvenient, and the water must be pumped out later, but it is considerably easier to do this than to replace a cracked or broken heavy concrete floor slab.

When the floor slab is poured in such cases, provisions should be made for deep keys where the foundation walls are to join it and for keys between the walls and their own footings. The design usually calls for reinforcing rods to be turned up into the walls and for the walls themselves to be tied together with reinforcing rods at the corners. The design will probably call for a porous pipe bedded in crushed stone and laid around the base of the walls. This is an excellent idea for any foundation where water may be encountered. The walls can be waterproofed by applying hot pitch and waterproof material on the outside, or by integral waterproofing mixed with the concrete, or by several layers of metallic waterproofing coating applied to the inside. Metallic waterproofing consists of small particles of iron which are mixed with cement and sand and water into a dense plaster. As the plaster dries and sets, the iron "rusts" and fills any interstices in the compound which then forms a watertight coating. The ingenuity of the architect engineer and his team of soil experts and geologists is equal to these challenges.

3. INSPECTION OF ADJOINING PROPERTY BEFORE WORK STARTS

3A. Use of the Seismograph

Surrounding property should be inspected before the excavation is commenced. As a matter of fact if there are buildings on the site to be razed before excavation, the inspection should be made before that operation. Where older buildings are concerned, a thorough inspection should be made of their exterior walls, their foundations walls, and generally of the condition of the interior. When noticeable cracking or other signs of settlement or age are found it is advisable to take photographs. Such inspection reports should be dated and signed and in come cases attested to. The author knows of several cases where photographs showing severe settlement cracks in an adjoining building before any new building operation was started saved expensive law suits and other harassment. If there is any reason to believe that there may be trouble it is advisable to engage an outside professional to make such inspections. A device that is now being used to record vibration caused by building operations in their "noisy" stages, such as building wrecking, blasting, and pile driving, is an adaptation of the seismograph. This instrument is portable and can record vibrations in three planes. It has been of great help in convincing public authorities that the operation going on is not endangering the public, or nearby streets, or property. It was of great help recently in a wrecking operation where several adjoining owners claimed that the large drop ball (headache ball) which was being used for breaking the concrete floors of an old warehouse, was causing excessive vibration which was cracking their building walls. Seismograph readings taken during the operation convinced the city authorities that this was not so. This prior inspection routine should be called to the attention of the owner by the architect engineer, who should advise him how and by whom it should be done.

CONTENTS

1. THE DETERMINATION OF THE FRAME TO USE
   A. Availability of Materials and Labor
   B. The Economics—Cost and Time
   C. Adaptation of Frame to the Architectural Concept

2. METHODS OF CONSTRUCTION OF VARIOUS STRUCTURAL FRAMES
   A. The Wood Frame
   B. The Wall Bearing or Light Steel Structure
   C. Structural Steel
      1. The basic design concept
      2. The typical specification
      3. The foundation
      4. The structure—A discussion of some outstanding variations
      5. The design systems most often used
      6. The connections
      7. Fire protection of the steel
   D. Structural Concrete
      1. The basic design
      2. The typical specification
      3. The foundation
      4. Design systems most often used
      5. Prestressed and post-tensioned members
   E. The Heavy Wood Structure

6

THE STRUCTURAL FRAME

1. THE DETERMINATION OF THE FRAME TO USE

The type of structural frame to be used for a project is determined by weighing many factors. The intended use for the structure determines the required clear distances between supporting walls or columns and the designed live load. A private residence may be designed for 12- to 20-foot clear spans between bearing supports and a two- or three-story office building may go up to a 25-foot clear span. As these structures must carry only a light live load, the design may call for a wall bearing[1] structure with interior bearing on light steel columns or masonry walls. If the structure is to be used as a manufacturing plant, the kind of machinery to be used and the requirement for column free space must be known. Depending on these it may become necessary to design a reinforced concrete or steel frame, even for a two- or three-story building. If the plant requires very large clear spaces on one floor, the designer may call for long span light steel or laminated wood roof supports bearing on masonry walls or steel columns. The laboratory or research building is usually in the same category as the small office building as far as the structural frame is concerned, but there may be exceptions. The use of heavy electric motors, steel retorts, complicated piping or highly combustible

---

[1] In a wall bearing structure, the exterior wall, which is usually of masonry, carries the weight of the floors. On the single-family residence the exterior wall and interior bearing partitions are usually of wood frame construction.

THE STRUCTURAL FRAME

material testing may necessitate heavy fireproof framing. In the research building there may be a need for complete soundproofing. In such cases the usual light steel frame may not do at all.

The small apartment building which must be fire-resistant, but not fireproof, may be built of masonry walls supporting light steel, or even wood floor beams, supported on interior masonry walls or light steel columns. Because the multistory office building or apartment building must be fire-resistant[2] and must meet rigid design standards set by code, the architect engineer has the choice of using reinforced concrete, structural steel, or a combination of concrete and steel as his structural frame. He can combine these with various floor systems and materials for his perimeter walls, but all such materials must meet exacting code requirements.

Many choices of material and method are available and the architect engineer, with the owner, must decide which structural frame will most closely fit his requirements for strength, economy, and the quality of the architecture. In an office building, flexibility of the structure is highly desirable so that its interior arrangement will not in any way inhibit the moving of functions or office partitions.

1A. Availability of Materials and Labor

Because there is a wide choice of materials the architect engineer must make several investigations before coming to a final decision. He should first investigate the availability of such materials and the availability of the skilled labor to use them properly. Of course, with the magnificent distribution system of this country, almost anything is available anywhere. The planner may use structural steel in the Ozarks or a heavy reinforced concrete structure in Pretty Prairie, Kansas, and his steel or cement and reinforcing rods will arrive safely and fairly quickly. But what about the labor to erect it? Away from large population centers, structural steel workers or skilled form carpenters and cement masons are hard to find. Perhaps the designer should consider heavy timber mill construction to carry heavy machinery in a low building. Even in large centers there are often severe shortages of skilled labor in certain trades. With a good amount of structural steel now being welded instead of riveted or bolted, there is apt to be a shortage of certified welders. The author knows of one project in a large city where, because no skilled men were available, the entire specification for the fireproofing of the steel structure by the use of forms and poured concrete had to be abandoned. The architect engineer was not familiar with local conditions.

[2]"Fire-resistant" means that a structure is so built that it can withstand certain high temperatures for a certain number of hours, without failure or collapse.

THE STRUCTURAL FRAME

1B. The Economics—Cost and Time

Next to be determined is the cost and time factor. Unless there is reason for a particular type of frame, it is simple for the architect engineer to call a contractor to discuss several alternates—for instance, for the small building, the use of a masonry wall supporting the second floor versus light steel columns and beams with an aluminum and glass curtain wall. For many structures the choice limits itself to reinforced concrete or structural steel, because of the structural strength required plus the fact that such buildings must be completely fireproof. The contenders generally state that reinforced concrete takes longer to erect but costs less to buy. This may be true for certain cases, but the owner and the architect engineer must weigh these factors carefully. Perhaps alternate schemes can be priced. The decision may be influenced by the difficulty of obtaining structural steel when it is required or the possibility of a carpenters' strike.

1C. Adaptation of Frame to the Architectural Concept

Last, the structural frame must fit the planner's and owner's concept of what the building is and does. The sizes and placing of columns to create clear space is very important. The creation of a modular system for partitions and lighting or the ease with which the structure may be enlarged must be considered. In difficult foundation situations the total weight of the structure may be important. Possibly the architect engineer wishes to create a light airy look or a solid conservative look, or he wants to show the "bones" of the structure as massive concrete, or masonry, or light steel. In many structures such as auditoriums, or monumental and institutional buildings, alternate schemes may be drawn and engineered in order to determine the type of frame that best suits the concept of both cost and aesthetics. It is well for all concerned to take a long hard look and to seek advice on costs and availability before a decision is made.

2. METHODS OF CONSTRUCTION OF VARIOUS STRUCTURAL FRAMES

All structural frames are subject to strict building code requirements. All communities having such codes have men who inspect plans and issue building permits. They vary from the small suburban town builder cum building inspector to the skilled structural engineer in the large city.

THE STRUCTURAL FRAME

2A. The Wood Frame

Because the wood frame is the simplest form of structure, it is the one most used for single family residences, small garden apartments, and other small structures. It is therefore the object of explicit requirements in small town codes. A typical code will state the maximum allowable clear span of floor joists, or rafters, or beams for certain sizes and certain spacing between them. (2 x 10 joists on 12-inch centers can span 15'0".) It will give allowable loading for posts and columns for specified unsupported lengths. The code may even recommend a nailing schedule. There are books, such as *Architectural Graphic Standards*, that show in complete detail the various ways of framing a house. With these aids and with further reference to the code concerning such matters as fire stopping, required size of lintels over opening, etc., the relatively inexperienced architect engineer can safely design a small structure.

It is sometimes advisable to combine wood framing with other materials. For a particularly long clear span it may be advisable to use a light steel I-beam supported on lally columns,[3] or the lintel beams over wide window or door openings may be steel angle iron. The supports under a girder beam which holds the floor joists could be lally columns or concrete masonry blocks.

There are some precautions which should be observed when using the wood frame. The wood sill on which it sits should be bolted to the foundation wall at intervals. The top course of concrete block should be filled in solidly with cement mortar to hold these bolts and to form a solid support for the sill; if a poured concrete foundation wall is being used, the bolts should be set in the form before the concrete is poured. The planner should visit the construction site often. Mechanics such as steamfitters, plumbers, and electricians drill holes through and hack away parts of structural members, *which makes it easier for them to work but which may later cause serious settling and cracking*. It is also suggested that when a structure is built in certain areas the planner specify termite shields and make sure that wood trim or other wood is not in direct contact with soil

2B. The Wall Bearing or Light Steel Structure

The more elaborate single-family residence, the small suburban business building, and the garden apartment building may use wood framing but the building codes discourage this. For a better fire-resistive rating than is possible for an all-wood frame, the planner can use gypsum exterior sheathing nailed to the 2 x 4 framing studs and an interior wall of type X or fire resistive gypsum wall board. This will give a one-hour rating which will do for a two-family house, or very small apartment or office building. Usually,

---

[3]This is a trade name for a steel pipe filled with concrete and used instead of a wood supporting post or a light steel column.

however, such buildings use a light steel frame or a structural quality masonry exterior wall. These support steel open web beams or bar joists on which the floor rests. Pleasing architectural combinations may be obtained by the use of masonry piers, separated by metal and glass spandrels and windows, in which the masonry serves the two-fold purpose of structural support and architectural concept. Most codes do not allow masonry bearing walls to be over 40 feet high and even then they must not be further than 25 feet from the nearest interior supporting member. The interior supports can be masonry walls, or steel columns, or lally columns which support light steel

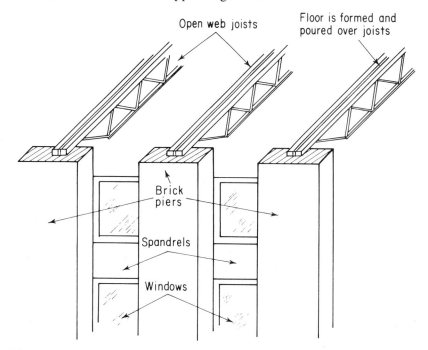

Masonry Wall Combined with Aluminum Windows and Aluminum or Glass Spandrels.

girder beams. The architect engineer's choice of materials depends (as previously mentioned) on the cost and availability of the material and labor, on his architectural design, and of course on the load the structure will be called upon to carry.

2C. STRUCTURAL STEEL

In Section 1 of this chapter there was a discussion of the uses of various structural frames. The trade organizations that represent the steel and the Portland cement industries have many things to say about the advantages of each material. Until recently tall buildings for office or housing use were, with few exceptions, built of structural steel. Some big city codes did not allow the use of anything but steel for a structure over a certain height. Be-

THE STRUCTURAL FRAME

cause of the many advances in concrete design, such codes are being changed. Structural steel, however, has certain advantages which the architect engineer must carefully consider. First, steel is quicker to erect. Any reasonably efficient erection crew can erect two stories a week in a normal high-rise job. Second, steel columns and beams are less bulky than concrete, although new concrete design is narrowing this gap. Third, a steel structure is lighter than one of reinforced concrete, which is an advantage in difficult foundation conditions. In an apartment house that does not require large clear space, a frame consisting of light structural steel for columns and girders and open web joists for the floor system can be quickly erected, and is reasonably priced. There are examples of such structures up to 19 stories in height. Generally, however, steel is more expensive and delivery may not be guaranteed to meet a schedule.

2C1. *The Basic Design Concept*

Now the architect engineer and owner after carefully weighing every factor, have decided on structural steel. The structural engineer has been told and he starts his design. He has the architect's preliminary plan and knows where the stairways and elevators are to be; where the duct shafts and riser shafts are to be and how much room they need; he knows the size of the module and the column bay size. He has been told the live loads that the structure is to carry, and the type of floor system and building skin, and other materials of construction that will constitute the weight of the structure itself. He designs a steel framework supported by columns and fastened by rivets, or bolts, or by welded joints, or by a combination of all three. He is the one who decides what the foundation shall be—whether pile or spread footing, and how large they are to be, and what strength they must develop. He furnishes a set of structural steel drawings for the foundation work and shows in detail how the structure is to bear on the foundation.

2C2. *The Typical Specification*

A typical specification mentions the following basic facts.

Erection. The frame shall be carried up true and plumb and be braced where necessary to take care of all loads, including the temporary load of the erection equipment itself. Exterior columns or elevator shaft columns may not be out of true more than 1 to 1000 (sometimes this is 1 to 1500), and beams may not be out of level more than 1 in 500 for their entire length. The specification for erection mentions temporary connections of sufficient strength to hold the structure. These connections are not to be removed until the structure is aligned, when they are replaced by permanent connections.

THE STRUCTURAL FRAME

Connections. The specification will mention high-strength bolting, riveting, and welding with a description of each method. It will tell how hot a rivet must be and how it should be driven; how hot surrounding metal must be before it can be welded and what kind of welding equipment may be used.

Testing. The specification will provide for the testing of steel, usually at the mill, or will accept steel mill tests which have been guaranteed. Some specifications simply quote ASTM specification numbers for the steel, the bolts, the rivets, the welding, etc. All specifications provide for the careful testing of welds by radiographic, ultrasonic, and other means. Because welding is a fairly new technique, the specification will call for certified welders[4] and constant inspection. The author knows of a building where careful testing by public authority after the structure was half-erected showed that a number of important structural welds were almost surface welds with no deep penetration because the welders were inexperienced, careless, or hurried. They had either not heated the joined members to the correct temperature, had carelessly handled the welding material, or were not given enough time to do a good job. There was a delay of over three months until all the welds were corrected. Meanwhile structural steel had to be stored and tenant rental income was lost. Building codes that are fairly administered are the ally of the conscientious architect engineer and owner.

2C3. *The Foundation*

The general character and use of the structure have now been determined. The following section will describe specific methods and materials of foundation design; of the steel design and of the various combinations possible for various uses. A steel structure rests on steel columns which bear on a foundation and the foundation must be so designed that the weight of the structure is distributed evenly on the bearing soil. After the concrete footing has been poured in accordance with the design, the structural designer specifies that a steel billet or heavy plate be set under the columns in order to spread the weight over the concrete footing. Such billets for a 20-story building may be 6'6" x 6'6" x 9" thick, and proportionately larger for heavier structures. They are usually set on wedges or leveling bolts and perfectly leveled before a cement grout (a thin gruel of cement and sand), is poured under them. The billets have heavy bolts welded to them and the structural steel columns are bolted down to them by means of angles. In some cases where the load is very heavy or the weight must be distributed over a larger area, a steel grillage is used under the steel billet. A grillage consists of a number of

---

[4]Such certificates are granted by a local authority after proper tests have been given under the rules of the American Welding Society. These certificates are proof of a welder's competence and can be carried by him from job to job.

THE STRUCTURAL FRAME

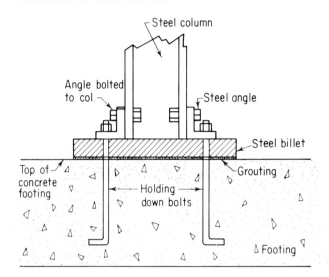

Steel Billet on Concrete Footing. The Entire Billet is Later Encased in a Concrete Floor.

steel I-beams held together by long bolts through their webs and separated by pipe collars. When a grillage is used it is completely encased in grout after it has been leveled and before a column is set on it.

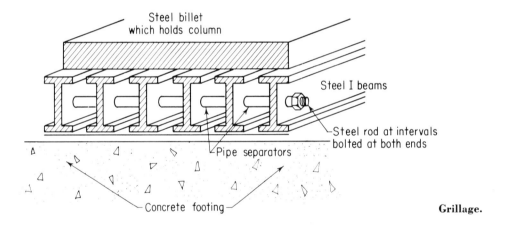

Grillage.

2C4. *The Structure—A Discussion of Some Outstanding Variations*

The size of the columns is determined by the weight of the structure, the live loads it is to bear, and the distance between columns (the size of the bay). Where seismic disturbances are apt to occur or where very heavy wind loads are possible, the determination of the column sizes takes this into consideration. Building codes generally describe the conditions that must be provided for. The advent of high-quality structural steels now makes it possible to preserve the same column size throughout the structure. The lower columns use the high-strength steel that can be changed to the normal structural

THE STRUCTURAL FRAME

steel (ASTM-A-36) at the proper height. The advantages of column uniformity are obvious. Where called for, the architect engineer must also provide for wind bracing. In the typical high-rise office building such wind bracing is usually placed at the core of the building where the strongly braced structural members surround elevator or stair shafts. Some recent structures have been designed for such structural bracing to be on the exterior of the building. The John Hancock Building in Chicago and the Alcoa Building in San Francisco are notable examples.

There are other possible variations of the structural frame. In the World Trade Center buildings in New York City the structural frame acts as the outer wall of the building. These are really steel, wall bearing structures. Some building designs bare the entire structural frame and use it as an architectural expression. In such cases the steel is very often of a type

The Alcoa Building, San Francisco, Calif. Skidmore, Owings & Merrill, Architects/Engineers. Photograph by Morley Baer.

John Hancock Center, Chicago, Ill. Skidmore, Owings & Merrill, Architects/Engineers.

THE STRUCTURAL FRAME

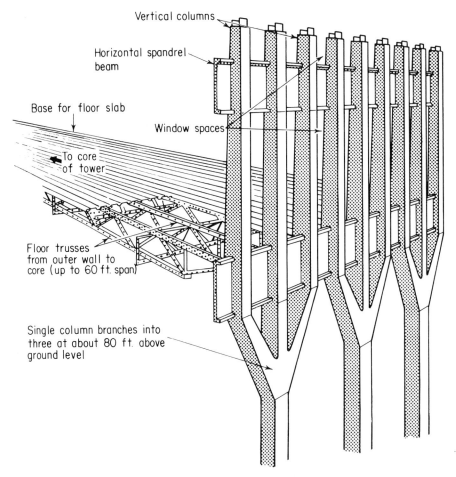

Outside walls' structural members, shown before getting sheathing, support half of tower's weight. (Tower's core takes other half.) Floor-to-ceiling windows go in 22-inch spaces between upper columns.

World Trade Center Framing. The Exterior Structural Wall Serves as a Building Skin and Structural Framing.

that will "rust" to a certain predetermined depth and this "rust" or oxidized coating forms a handsome protection against further weathering. Some buildings have been designed to taper up from the base like the Eiffel Tower. Notable examples are the John Hancock Building and the First National Bank Building in Chicago. The new United States Steel Building in Pittsburgh is an outstanding example of the ingenuity of the architect engineer team. This is a triangular shaped building 54 stories high with column free interior spans of 45′6″. The building frame consists of structural columns placed on the

THE STRUCTURAL FRAME

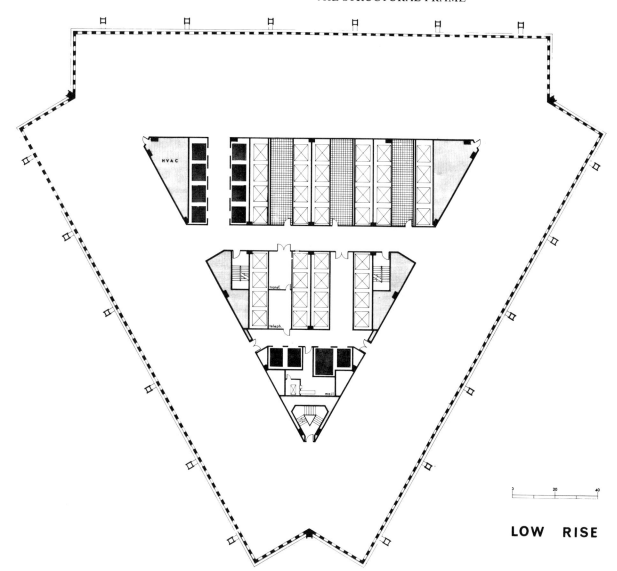

LOW RISE

U.S. Steel Building, Pittsburgh, Pa. Harrison, Abromowitz & Abbe Architects. Worthington, Skilling, Helle & Jackson —Edwards & Hjorth—Associated Structural Engineers.

exterior of the building skin with steel girders spanning the interior to the columns around the central utility core. These exterior columns are *not* fireproofed and in order to comply with the code they are filled with water which can circulate and protect them against the normal expansion and contraction caused by weather, and against extraordinary heat caused by fire. Heavy floor girders are located at every third floor. The two floors in between are held by comparatively light steel framing resting on these girders. And so on— with the new high strength steels; with the constant advance in design and

83

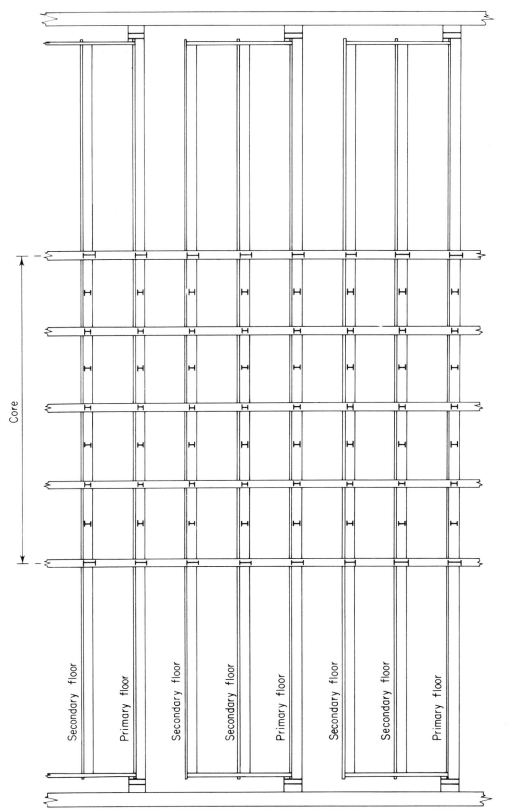

U. S. Steel Building, Pittsburg, Pa. Typical Cross Section.

with the recognition of this by the building codes, there is almost no limit to what can be done with the structural frame. The surface has barely been scratched and the field is wide open for the ingenious designer.

2C5. *The Design Systems Most Often Used*

There are three steel design systems now being used—"plastic," "composite," and "elastic." "Plastic" design, which at this time has limited use, allows the steel to bear loads beyond its certified strength by using its reserve strength and ductility to bear loads beyond its normal yield point. This is a very complicated and comparatively new theory and has not been accepted[5] beyond low simple structures or beams between normally designed girders. The "composite" design is widely used and accepted. Very simply, it recognizes that the structural frame and the structural floor can, if properly connected, act as a unit. A concrete floor connected to a steel beam by means of steel studs welded to the top of the beam and surrounded by concrete can act as a huge T-beam which resists compression as a unit. The floor and steel acting together can therefore each be lighter in weight. The "elastic" design takes cognizance of the varying strengths of different grades of steel and allows the designer to use such strengths to advantage. Before the acceptance of this design the engineer always had to design to the strength of standard steel A 36 whose minimum yield point is 36,000 psi.

2C6. *The Connections*

With these designs the engineer must specify the connections. Steel can be connected by rivets, high-strength bolts, or welding. The designer usually specifies a combination of these depending on whether he requires a perfectly rigid connection or one which allows some movement between members. The manner in which these connections are to be made and the characteristics and strength of the materials is carefully specified and in any case is set by building codes. The AISC and the ASTM publish complete detailed information on this subject.

2C7. *Fire Protection of the Steel*

An important part of the design of the structural steel frame is its protection against fire or intense heat. It may seem odd to the uninitiated to have to protect steel against heat. A glance at the photograph showing what happened to unprotected steel in a fire will prove conclusive. Structural shapes are rolled from billets of almost molten steel and are stretched, and

---

[5]There is one medium-rise apartment house now built by this design and no doubt others will follow.

THE STRUCTURAL FRAME

Unprotected Structural Steel After a Fire.

pushed, and pulled into desired shapes. In the structure they are in shear, or tension, or compression. This is how they do their job. When such members are subjected to enough heat to make them plastic, however, they all try to go their own way and the results can be incredible. All codes and the AISC rules are very explicit about the protection of steel against heat. Depending on the structure such rules will call for up to a four-hour rating for columns and up to three hours for beams. There are many ways of protecting the steel. These methods vary from the simple fire-resistant ceiling and the fire-resistant flooring surrounding a lightweight bar joist to the poured concrete fireproofing around beams and columns in a large structure. Following is a list of a few methods:

1. A column can be protected by fire-resistant gypsum wall board that is attached by adhesive and tied by wire. This is acceptable for lightweight steel or lally columns.
2. A column can be wrapped in wire lath which is plastered with coats of heat-resistant plaster (Vermiculite or Perlite) to the specified thickness.
3. A column can be surrounded by gypsum block which in turn is plastered over.

THE STRUCTURAL FRAME

4. It can be coated with several inches of mineral fiber which is mixed with an adhesive and sprayed on. This method is being used more and more extensively. It cannot be used, however, except where it is protected against rubbing or abrasion.
5. Lightweight floor beams can be protected by gypsum board fire-resistant ceilings.
6. Heavier beams can be protected by concrete which is poured into forms to surround the beam. This concrete usually forms an integral part of the floor system.
7. Such beams can also be heat-proofed by sprayed on mineral fiber fireproofing which is used today in most high-rise office building structures.

2D. STRUCTURAL CONCRETE

In the previous section, a number of reasons were given for the architect engineer's decision to use structural steel. As many reasons can be given for a decision in favor of structural concrete. New design concepts have succeeded in decreasing the sizes of the structural members. The design of the concrete structure now follows its own concepts and no longer imitates the typical steel design of column, girder, and beam. Concrete does not have to be fireproofed; it is generally less expensive, and the plasticity of its very nature lends itself to the free flowing or boldly massive architectural concept. There are many excellent examples, two of which are shown here. Until recently the European countries were far ahead of the United States in concrete design. This was because in Europe structural steel was hard to get

Alley Theatre. Photograph by Ezra Stoller (ESTO). Ulrich Franzen & Associates, Architects; Weiskopf & Pickworth, Structural Engineers.

THE STRUCTURAL FRAME

University of Illinois Assembly Hall. Harrison & Abramovitz, Architects.

and very expensive, and because the design of the structure required a great deal more engineering time, which was reasonably priced in Europe but prohibitive in this country. Another reason is that the labor required to erect the complicated forms necessary for intricate design is reasonably priced in Europe. Our domestic concrete design, on the other hand, except in monumental buildings, confines itself to fairly simple and repetitive patterns that require a minimum of labor. The computer has also made a difference. A computer can now give the answers in a day or two (the programming takes the time) that formerly took a dozen engineers several weeks. However, complicated form work is still expensive. The economic consideration can never be ignored and the intelligent architect engineer will be well advised to keep this in mind. However, this should not keep him from designing aesthetically pleasing structures.

2D1. *The Basic Design*

As in steel design, the structural designer is given the same information as to the clear spans required, the clear floor heights, the size and shape of the utility core, and the exterior design. He designs a structural concrete

THE STRUCTURAL FRAME

University of Illinois Assembly Hall. Harrison & Abramovitz, Architects.

framework that will meet these requirements. Because of the latitude now allowed by codes, he has many choices as to how the members will interact and what the floor and wall system will be. Instead of using structural members that are hoisted into place and fastened together, the structural designer must think of the structure as a single entity acting together. His structure must be poured in place, a piece at a time, and his columns, and floors, and walls must be connected by means of steel reinforcement which holds it all together and transmits stresses from one member to another.

2D2. *The Typical Specification*

A typical specification for a reinforced concrete structural frame will mention forms, reinforcing, testing, and weather.

Forms. To rigidly and substantially construct forms to be erected to line, shape, and dimension, and in precise position to form the lines and designs indicated. The forms, depending on the purpose and exposure of the concrete, may be of wooden boards or plywood, millboard, metal or fiber glass. These forms, prior to the pouring of concrete, must be substantially shored and braced (to hold the dead weight of the concrete until it can hold itself). Some of this shoring must be kept in place for as long as 28 days or until the full strength of the concrete has been developed.

Reinforcing. The reinforcing steel shall be of deformed bars (the steel reinforcing rods have bumps on them), conforming to specified ASTM standards of tensile strength. The tensile strength of the wire mesh should also conform to ASTM standards. The steel mill analysis may be accepted if it is properly documented. The electrodes for welding the steel must conform to American Welding Society or ASTM standards. Reinforcing bars must maintain the proper distances from one another and from the form. Such bars must be held in place by spreaders, or spacers, or chairs.[6] No splices may be made at any points of maximum stress unless so indicated where one bar will lap over another to develop more strength.

Concrete. The kind of aggregate[7] is specified according to size and character. (Per cent passing by weight for 1½ inch aggregate—95 to 100 through a 1½ inch sieve, 75 to 85 through a 1 inch sieve and so on. Clean, hard, fine-grained stone.) A "designed" mix of water, cement, sand, and stone is specified by the structural engineer. This mix is calculated to produce the necessary strength. It is specified that ready mix concrete (if used), shall be mixed almost dry in designed proportions until it arrives at the site, after which a proper additional amount of water is added. The concrete must be so placed that it does not separate—i.e., the heavy ingredients should not drop to the bottom. The same specifications hold for concrete mixed on the job. This occurs often in large construction projects.

Testing. The testing agency[8] must constantly check at the mixing plant to determine that the correct amount of each ingredient is used. An inspector must be at hand when the concrete is delivered to the site and must take test cylinders[9] and slump test[10] the mixture. He must watch constantly to see that the concrete is placed properly in the forms so that it does not separate and that it properly covers the reinforcing steel.

[6] A chair is a metal device that holds a reinforcing bar up from the bottom of the form. The bar sits on it.
[7] The aggregate is the stone or gravel, and the cement, and sand, and water that are all mixed to make concrete.
[8] A testing agency is always employed to check quality and strength of both concrete and steel. The reports of such agencies (when they are highly qualified) are accepted by the authorities.
[9] A test cylinder is a heavy cardboard cylinder which is filled with concrete from each batch and stored under specified temperature and humidity, some for 7 days and some for 28 days. The concrete within the cylinder is then tested for the specified compressive strength and a certificate is issued.
[10] Slump testing is used to determine the plasticity of the concrete. A metal truncated cone, 12-inches high, is filled with wet concrete. When the cone is lifted the wet concrete "slumps". About 4-inch slump is the usual allowance for normal work.

THE STRUCTURAL FRAME

Weather. There are prohibitions against placing concrete in freezing or rainy weather, especially in open forms. There are specified additives to use for protection against freezing. There are specified regulations for protecting the concrete against too rapid drying. (We have all seen concrete highways covered with wet burlap or heavy paper or hay.) If concrete loses its water content too quickly it becomes brittle.

2D3. *The Foundation*

The foundation for the concrete structure is no different from any other foundation. It may be on piles, on a footing on rock, or on spread footings. The footing sizes depend on the weight to be placed on them, and the concrete columns are poured over the footings and connected to them by dowels of reinforcing steel that have been set in the footings and project into the columns. Unlike steel, the concrete column requires no device to spread its weight because it is of sufficient size to perform this function itself.

2D4. *Design Systems Most Often Used*

As mentioned before, the design of the reinforced concrete structural frame has gone its own way and has taken advantage of advances in designed concrete mixes that without question can now attain the specified strength. The designer also takes advantage of the complex forms into which concrete can be poured and through which it can perform not only structural functions but can serve as a conduit for air or electricity or can become an architectural exterior.

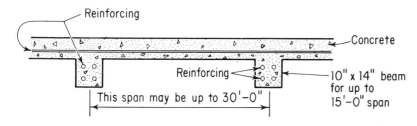

Section Of Beam & Slab Construction
In Reinforced Concrete

One of the simplest designs is the beam and slab design that actually follows ordinary steel design but uses concrete beams that are cast integrally with the floor slabs. This system is used in smaller office buildings and sometimes used in apartment buildings, or other structures where the beams are not objectionable and can be hidden by partitions directly under them. Where the clear spans are small (up to 15 feet), the beams can be as small as 10 inches wide by 14 inches deep but the size can go up to 18 inches deep by 8 feet wide, for instance, in a garage with a 30-foot clear span. As can be seen

THE STRUCTURAL FRAME

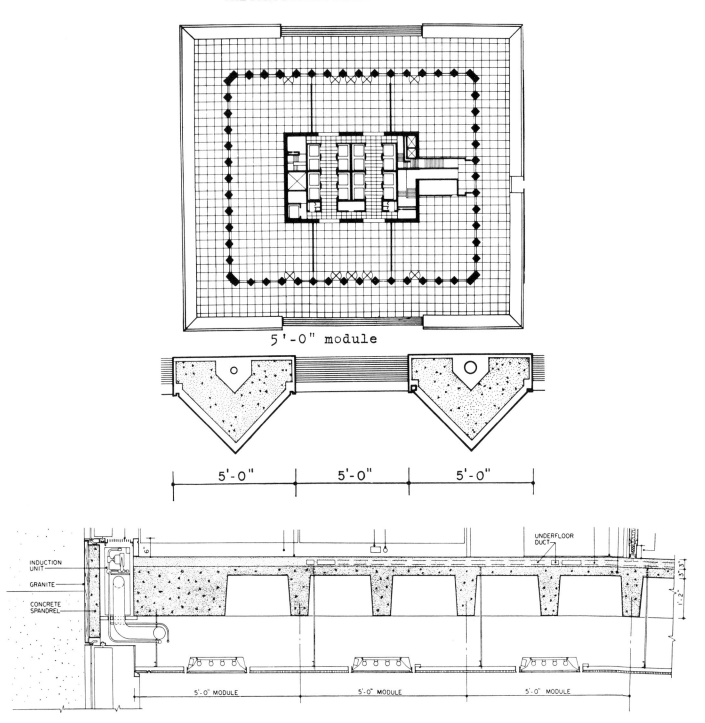

C.B.S. Building New York, N.Y. Top: Plan Showing Exterior Column Section. Bottom: Section Showing Waffle Floor. Eero Saarinen, Architect; Paul Weidlinger, Structural Engineer.

THE STRUCTURAL FRAME

in the sketch, the reinforcement is simple and the forms can be used over and over. This produces an economical structure. In tall buildings that use this system, the structural designer has to provide for wind bracing by heavy bands of reinforcement and concrete at the center core. This design produces a useful, economically viable structure.

The flat plate design was first used in an apartment building about 1944. It was an answer to short span structural steel that had to be protected with concrete fireproofing and where the beams projected down from every ceiling. There was also at that time a serious shortage of structural steel. The flat plate utilized bent reinforcing bars that projected at right angles and in two directions from every column, and supported flat slabs spanning 12 or 15 feet in both directions. This seemed like an economical solution of the use of concrete in apartment buildings. However, one of the economies involved the direct plastering of the underside of the slab, and the plaster would not bond to the smooth concrete. Another disadvantage was that the forest of bent reinforcing bars interfered with the placing of electric conduits. Some of the problems have since been solved and variations of the flat plate are often used.

The CBS Building in New York City is an outstanding example of what can be done with architectural concrete frames. This 38-story building is supported by its outside columns and its inside solid concrete core walls. The outside columns are triangular in shape and are each 5'0" less wide. The space, between each column, that is almost completely filled in with glass is also 5'0' wide. The waffle floor[11] is built to a 5'0' module with every other rib on 5'0" center. The total depth of the ribs and the structural slab is 1'5". The hung ceiling below distributes the air through the lighting fixtures and electricity is distributed by underfloor ducts above the structural slab. The clear span is 35' 0" all around the central core.

Another excellent application of long span construction using the waffle floor, combined with architectural poured exterior columns and walls and the interior mechanical and electrical system to form a complete building system, is L'Enfant Plaza, an office building development in Washington, D.C. This monumentally designed, but extremely practical, structure is framed with deep concrete ribs that form rectangles 3' 1" by 6' 2" by 1' 6¾" deep. They form bays that are 27' 9" long by 22' 0" wide. The floors, the exterior columns, and the interior shear or bracing walls are all hollow and are used for air distribution. The spaces between the ribs are used for the installation of the lighting fixtures that also serve to distribute the air which comes through the floor. The structure therefore becomes the air distribution system. The coffered ceiling becomes the lighting system. Because the con-

---

[11] A waffle floor looks like a waffle from underneath and is of course smooth on top.

THE STRUCTURAL FRAME

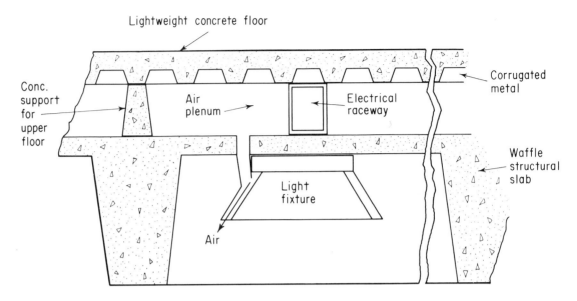

L'Enfant Plaza. Section through Floor. I.M. Pei & Partners, Architects; Weiskopf & Pickworth, Structural Engineers.

crete was cast in fiber glass coated forms, it is smooth and becomes the finished ceiling. The complete structural depth of ceiling to floor (2′8″) is considerably less than the space that would be taken by a conventional hung ceiling in which air ducts, electrical conduits, and lighting fixtures have to be placed. Even the exterior columns are poured around hollow ducts which serve as vertical air distributors. There is a variation of the waffle floor known as the one way pan that is a variation of the beam and slab. It acts as a T-bar and can cover spans up to 30 feet.

2D5. *Prestressed and Post-Tensioned Members*

Reinforced concrete reaches its highest possible strength when it is used in prestressed or post-tensioned members. Spans as long as 100 feet can be attained in members as shallow as three feet deep for roof loads. Such members are used everywhere for highway bridges and for the roof support from banks to manufacturing plants. These structural shapes have wide use in other spaces requiring long clear spans, such as office buildings or even garages where too many columns can infringe on parking space. The principle involved is simple. If one holds a number of books or blocks together loosely, the middle ones will fall. The tighter they are held together the more objects can be held. If such objects are strung on a flexible wire they will sag. If the wire is tightened they will assume a horizontal line and the stronger and tighter the wire, the longer the clear unsupported horizontal

THE STRUCTURAL FRAME

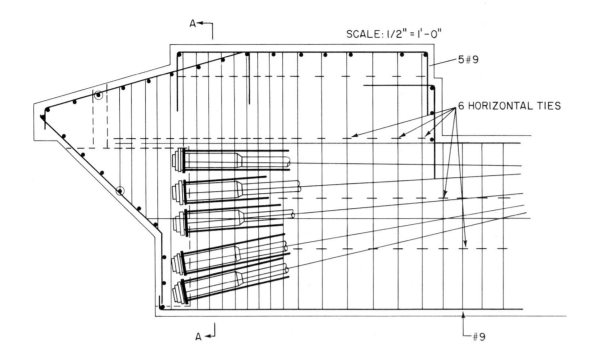

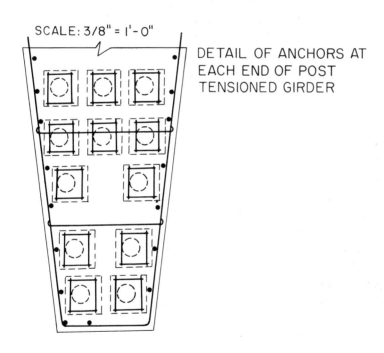

Detail of Anchors at Each End of Post-tensioned Girder. Below: Section A-A Typical for All Girders. Weiskopf & Pickworth, Structural Engineers.

THE STRUCTURAL FRAME

span. The tightened wire holds the objects together so closely that their surface friction prevents slippage. In prestressed (pretensioned) members, reinforcing rods or high tensile strength wires are stretched to a certain determined limit and then high-strength concrete (5000–6000 psi) is poured around them. When the concrete has set and holds the steel in a tight grip the stretched steel tries to return to its original length, but can't. It holds the concrete in such a tight grip that the tremendous friction between the reinforcing and the components of the concrete prevent any slippage or sagging, and the steel can thus sustain very heavy weight. In post-tensioning, cables are left loose in hollow tubes that have been set at certain intervals in the form before the concrete is poured. These wires are permanently fastened at one end. After the concrete has set, and the structural member is on the job, the cables are stretched tight by powerful jacks and anchored securely into place by end plates to which the cables are fastened, and by grouting around the cables. This post-tensioning acts in the same manner as the pre-tensioning. The use of this form of structural concrete is being constantly expanded.

2E. THE HEAVY WOOD STRUCTURE

In many instances wood can be used for the structural frame. In a plant where there is exposure to corrosive vapors, it is found that wood has much more resistance to chemical action than does steel. The heavy wood structural frame is used extensively in warehouses or manufacturing plants where nonflammable materials are stored or manufactured and where the labor and material for concrete or steel framing is difficult to obtain. The heavy post and beam frame which holds up two- or three-ply laminated wood floors produces a very satisfactory structure and should be carefully considered for certain uses and areas.

An important use of wood is in the laminated structural arch or truss. The arch is made of several thicknesses of wood glued together and shaped into the form required. Places of public assembly such as churches, bowling alleys, gymnasiums, and swimming pools use these arches or trusses extensively. They come in all shapes and sizes and can be built to span up to 120 feet. A bow string truss that spans such a length is about 17 feet high at its center. A laminated arch to be used for a church might be 18 inches wide by 78 inches deep and span 100 feet. These sizes are given to show the extent of the development and use of this material. One other advantage of the wood arch is that its resistance to heat (believe it or not) is better than that of steel. At a temperature of 1100°, structural steel loses half of its tensile strength and at 1700° it cannot support its own dead weight. The laminated arch or truss, however, burns slowly and loses its strength only as it burns. Standard fire tests show that at standard fire temperatures a heavy wood mem-

THE STRUCTURAL FRAME

Examples of Heavy Laminated Timber Construction.

THE STRUCTURAL FRAME

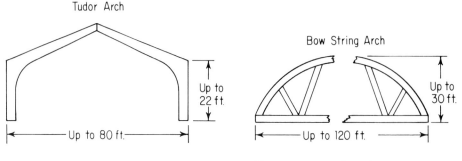

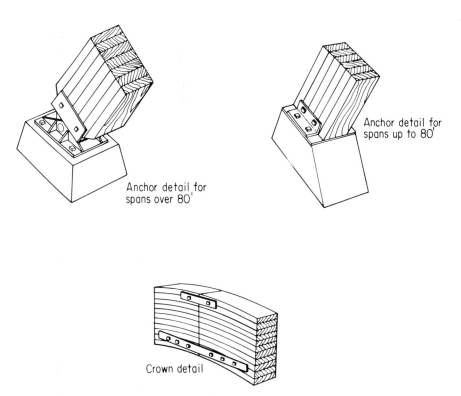

Glued Laminated Wood Structural Members.

ber burns at a rate of 1 inch in 33 minutes. This characteristic of slow burning and slow loss of strength is very important. Recently, a fire in a large convention hall created a total loss because the steel roof trusses twisted under the heat and practically pulled down the entire structure. Wood trusses would simply have burned through and collapsed leaving the exterior walls standing.

There are other kinds of structural systems that are a composite of many that have been mentioned here, but the great majority of structures are built using one of the described methods. The ingenious architect engineer can always improvise or invent to suit his particular needs. He need only be fully cognizant of engineering and architectural principles and of the requirements of the various codes.

CONTENTS

1. THE DETERMINATION OF THE SYSTEM AND MATERIALS TO BE USED

2. THE LOW LIGHT USE STRUCTURE
   A. Description of Various Flooring Materials

3. THE STRUCTURAL STEEL BUILDING
   A. Reinforced Concrete Floor
   B. The Cellular or Corrugated Steel Floor
      1. Typical Specification
      2. The Types Available and Their Use

4. THE REINFORCED CONCRETE BUILDING
   A. The Adaptability of the Floor

7

THE STRUCTURAL FLOOR

1. THE DETERMINATION OF THE SYSTEM AND MATERIALS TO BE USED

The structural floor, like the structural frame, is not usually seen. It is under the finished floor and is an integral part of the structural frame. The structural floor has a very definite job to do. It acts in concert with the frame whether it be steel, wood, or concrete, to stiffen the structure and to help the frame resist tension, compression, and shear. The choice of the floor system depends to some extent on the material of the structural frame, but there are always alternatives. The material of the floor and the method of its installation depend on the load it is to bear and on the use to which it will be put. The architect engineer should ask a number of questions before finally deciding on a floor system. First, he must be sure that his floor system will be compatible with his structural frame. Second, he must choose the floor that will be capable of bearing the designed loads. Third, he must be sure that his floor will be of sufficient strength to stiffen the structure against wind stresses, seismic disturbance, or the vibration of heavy equipment. In this connection also, he must determine that the method of fastening the floor to the structure will be capable of distributing such forces between them. Fourth, he must be sure that the floor has the proper fire-resistance. Fifth, he should use a floor system that lends itself to the proper distribution of necessary building utilities. He must also choose the floor that produces the greatest strength for the least thickness and he certainly must investigate

THE STRUCTURAL FLOOR

the cost of his various possible choices and their availability. Even in wood framed private residences, the wooden underfloor, whether it be of plywood or pine boards, stiffens them against sway when it is nailed to the floor joists. It also stiffens them against compression, by acting as a continuous T member and it helps with the shear load at the bearing points.

2. THE LOW LIGHT USE STRUCTURE

In the low light use structure such as an office building, or research or medical building, or in the garden apartment, the variety of available floor materials ranges from double laminated plywood to concrete plank, or light corrugated steel.

2A. DESCRIPTION OF VARIOUS FLOORING MATERIALS

Double laminated plywood can be covered by carpeting, or finished wood, or resilient tile floor which is often used. It is used over either wood joists or expanded steel joists that have wood nailer strips attached to them.

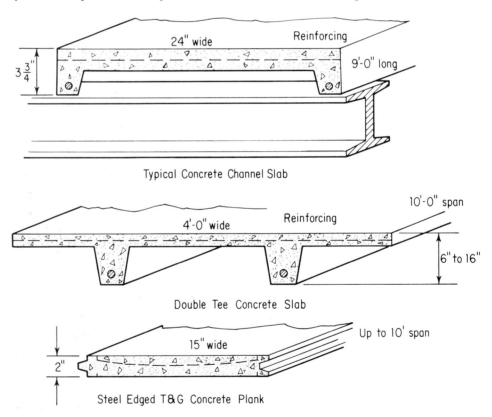

Channel, Double Tee, and T&G.

THE STRUCTURAL FLOOR

This type of flooring can attain a fire rating by the use of a fire retardant ceiling below. Concrete plank that comes in various standard sizes and shapes and is usually reinforced with light steel mesh can also be used. Such plank can be obtained as a steel edged shape. It is tongue and grooved on all four sides and a 2-inch thick plank can carry over 100 pounds per square foot on a 5-foot span and will have a 2-hour fire rating. This makes it acceptable for low multifamily dwellings or small office buildings. There are variants of this such as the concrete channel slab that is shaped as shown and contains a reinforcing rod in each protrusion; or for larger structures the double tee slab that can carry a 100-pound per square foot floor load on a 10-foot span. Such precast concrete floor slabs are fastened to the steel joists by clips so that they can act with the steel. They are simple to erect with labor that need not be highly skilled. The underside of the tee slab can be used as a channel for wiring or air. When these slabs are covered with a coating of cement finish they form a base for any finished floor covering. In instances where a wood floor is required, the concrete plank will support wood sleepers (furring strips), to which a finished wood floor can be nailed. All of these floors can be used in conjunction with steel bar joists, light steel I-beams, or concrete beams.

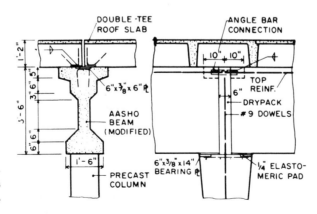

Double Tee Beams on Precast I-Beams (Top). I-Beams Welded Connection (Below).

THE STRUCTURAL FLOOR

There is also available a light steel corrugated floor that may be spotwelded to the steel bar joists and then covered with concrete, usually 1½ inches thick. In a medical or research building where there may be use for numerous electrical outlets in the floor, the corrugations can be made large enough so that any chosen number can be used for enclosing the wire to provide floor outlets at any place that they are needed. The concrete floor is then covered with any chosen finished floor. Such a floor in conjunction with a fire-retardant ceiling underneath attains a maximum fire rating. The concrete also acts as a structural stiffener. A floor of this kind can serve many purposes, is economical, and should be looked into carefully.

3. THE STRUCTURAL STEEL BUILDING

The structural steel building gives the architect engineer a choice of two basic floor systems which can be used separately or in combination.

3A. Reinforced Concrete Floor

In structural steel buildings, until very recently, the entire floor system, with few exceptions, consisted of reinforced concrete that was poured over and around the structural steel beams and columns, and served not only as fireproofing but as part of the structural system. While this is still used to some extent for an entire building, the cost of labor for forming around the beams, under the slabs, and around the columns has become almost prohibitive and the supply of skilled labor that can perform this work has dwindled. The concrete floor slab is now used where heavy floor loading is required. In a building which may use a corrugated or cellular subfloor for all normally used areas, the machine or equipment areas with their heavy loading and vibration must rest on reinforced concrete. The specifications for the erection of such floor slabs follows the specifications used for the structural concrete frame. The placing of forms, the materials to be used and their mixing and testing, the careful placing of the reinforcement are all spelled out in detail and show the importance that the architect engineer attaches to this floor system.

In the past several years the structural engineer has started to design concrete slabs to act in conjunction with the steel, not by surrounding the structural beams with concrete but by positively fastening the concrete to the top of the beam by means of steel studs welded to the steel and imbedded in the concrete. This has been mentioned in Chapter 6 on the structural frame. Such composite floors save material and labor, and can be used for either the complete concrete slab or for a combination of steel decking and poured concrete finish floor. The sketch shows a typical cross-section of such a floor. It can be seen that the concrete floor is serving to stiffen the entire

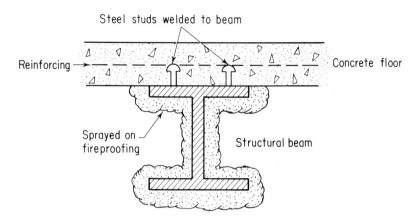

Composite Design. The Welded Studs Act with the Reinforced Concrete Floor to Form a Tee Beam which Stiffens the Entire Structure.

structural steel building system. As a matter of note, most building codes prohibit structural steel from being erected more than four or five stories above the structural floor system.

3B. The Cellular or Corrugated Steel Floor

In the last few years the corrugated steel floor has come more and more into use. Such flooring, which can be hoisted to follow closely behind structural steel, can be placed loosely over the structural steel beams and therefore provide a surface for men to work on. The corrugated steel floor comes in many forms and can be used to perform many functions. It can carry high and low-tension electricity and it can carry air. It can be obtained in shapes where it attaches itself to the concrete that must be poured over it, so as to form a complete structural member in conjunction with the supporting beams.

3B1. *Typical Specification*

A typical specification for the use of such a floor would be as follows.

Materials. The type of steel is specified to conform to ASTM standards and the manner of its forming into the proper shapes is referred to in the standards of the American Iron and Steel Institute and the Metal Roof Deck Technical Institute. (As mentioned previously the various trade organizations have issued standard specifications for their materials that are accepted to a great extent by public authorities and serve to protect the owner against shoddy, substandard materials.) A zinc coating is usually specified. The size of the corrugation and the gauge of the steel is referred to in the drawings.

THE STRUCTURAL FLOOR

Erection. The units must be placed so as to bear firmly on the structural steel members (2½-inch minimum end bearing and 1½-inch minimum side bearing). The fluting or corrugations must line up exactly with each other from sheet to sheet. The sheets must be fastened securely to the steel framework by welds that are not less than ¾ inch in diameter and not more than 12 inches apart across their width. Abutting sheets must be fastened together at not less than 3-foot intervals.

3B2. *Types of Steel Deck Floors Available and Their Use*

The steel deck floor can perform its functions in several ways. It can act simply as a form for concrete in a variety of "light use buildings." It can act as a form for concrete and, as previously, described, also perform a structural function in combination with the concrete in a composite floor.

The most universally used steel deck is the cellular floor section which is between 1½ and 3 inches deep and in which certain cells are covered below by sheet metal to form channels for electric wiring. This flooring comes in various lengths but the most economical use is for a clear span of 10 feet or less. For instance, for 10-foot clear span, a 3-inch deep floor made of 18 gauge steel will carry a live load of 50 pounds per square foot. This flooring has tabs underneath it so that metal hangers can be attached to hold the ceiling below. This floor is always covered with concrete either as a simple floor, or as a composite floor in which case the concrete is reinforced. This flooring can also be made to carry air. In this case certain cells are made wider than the normal. These cells come in pairs—one to carry warm air and one to carry cold. The cells usually end at a mixing unit at the perimeter of the structure that acts to air condition the perimeter area. The floor in this case eliminates the usual sheet metal ducts that are used to convey air from the central core area to the perimeter. The flooring can also be deformed or crimped in such a way that it acts with a reinforced concrete floor as a composite floor without the use of studs welded to the beams. All steel decks attain their greatest fire-resistance when they are sprayed on their underside with mineral fiber to a specified depth, depending on the depth of the floor and its gauge. A three-hour fire rating is usually required. A lesser fire rating for smaller buildings can be attained by the use of a fire-resistant ceiling below. Tests have shown that metal deck when covered with light-weight concrete, has a two-hour fire rating without spray fire proofing.

4. THE REINFORCED CONCRETE BUILDING

In Chapter 6, Section 2D, there were descriptions of the various floor systems that could be used. These floor systems in all structural concrete buildings are an integral part of the structure and are poured in place and are inex-

THE STRUCTURAL FLOOR

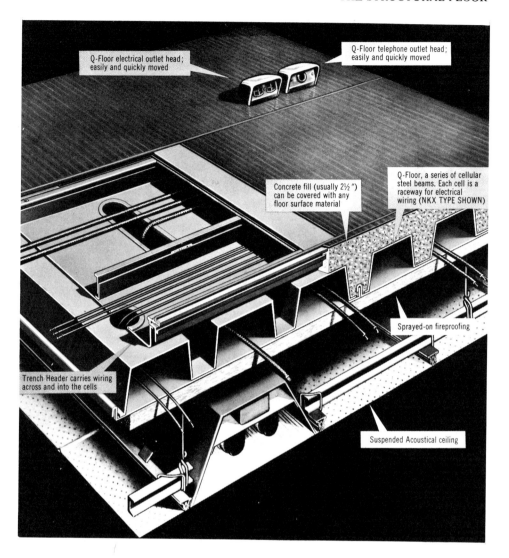

Robertson Q Floor.

tricably connected with the columns, the beams, and each other. Obviously they cannot be hoisted into place later and rearranged to suit some purpose not originally planned for.

4A. THE ADAPTABILITY OF THE FLOOR

These floor systems can, however, be adapted to serve purposes other than their structural use. The long pan forms or the waffle floors can be adapted to carry both electrical conduit and air within the depth of the floor. The waffle floor may have holes through the fins in one or two directions to

THE STRUCTURAL FLOOR

carry air. Where air is not wanted these holes can be plugged but can be opened when needed. In one case the structural floor carries an overfloor of concrete that is cast on a metal form and forms cells between the ribs that support it. These cells carry conditioned air that is borne down through the structural floor and is distributed by means of lighting fixtures situated within the ribs of the waffle floor below (see sketch of L'Enfant Plaza floor system on page 94). Many structures that use prestressed concrete beams use spaces between the ribs of these beams to carry electrical and air ducts.

Research by manufacturers of the various types of flooring, and advanced research in concrete design and their acceptance by the various building codes, presents an ever present challenge to the designer.

CONTENTS

1. DETERMINATION OF THE TOTAL ELECTRICAL REQUIREMENTS AND THE DISTRIBUTION
   A. The Residence
   B. The Small Business Building
   C. The Multistory Business or Residence Building
      1. A specification on requirements
      2. Occupancy requirements
      3. Equipment requirements
      4. The distribution

2. TYPICAL POWER COMPANY DISTRIBUTION

3. SELECTION OF EQUIPMENT

4. LIGHTING

5. EMERGENCY GENERATORS

8

THE ELECTRICAL INSTALLATION

1. DETERMINATION OF THE TOTAL ELECTRICAL REQUIREMENTS AND THE DISTRIBUTION

In the preceding chapters the author has discussed the steps that must be taken by the architect engineer team in order to create a structure. Through the combined and applied knowledge of this team the structure has been placed in the correct location on a suitable site. The structure complies with the zoning and building regulations; the plans and specifications are completed; and a building permit has been issued. A contractor has been employed and he, the owner and the architect engineer know their respective duties and relationship to one another. The foundation has been completed and the structural frame and the structural floor system have been chosen and are being installed. Now this empty wood frame or masonry shell, or mass of steel and concrete must be given life, light, and warmth. There must also be facilities for the preservation and preparation of food. The business building must be provided with facilities to carry on its business, whether it be an adding machine, an X-ray machine, or a dentist's drill. Electricity is the answer to all these needs, and how it is provided and distributed is the subject of this chapter.

1A. The Residence

Before starting his design, the planner must have a knowledge of the requirements. This is a relatively simple process for the private residence

THE ELECTRICAL INSTALLATION

but even in this case the use of electricity is constantly increasing. In recent years almost every home needs provisions for a washing machine, dryer, dishwasher, refrigerator, toaster, electric oven, hot water heater, air conditioners (window units or central units) and the end is not yet in sight. Most codes state that in the event that any change is made in an electrical system that does not now provide for all these uses, plus future expansion, that an entirely new service must be installed. Codes also provide that the public utility feeder lines to the house be of sufficient capacity to accommodate possible future expanded use. The architect designing such a home would be wise to speak to a good local electrical contractor as well as the local building inspector and the power company to make sure that he is specifying adequate service and circuit wiring. He should be careful not to put too many heavy-use devices on the same circuit. There must also be provisions for a 230-volt service for air conditioning, or an electric stove, or a water heater. The house service in a small town is usually a 230-volt, three-wire system which can provide 115 volts or 230 volts as required. This service comes through a property box and a main switch, and fuses into a panel box that distributes the current to a series of circuits through circuit breakers, some of 15-ampere capacity and some of 20-ampere capacity. The circuit breakers are often used instead of fuses and trip open at overload. The usual practice is to have a minimum of 8 to 10 outlets per circuit. There are separate circuits for the

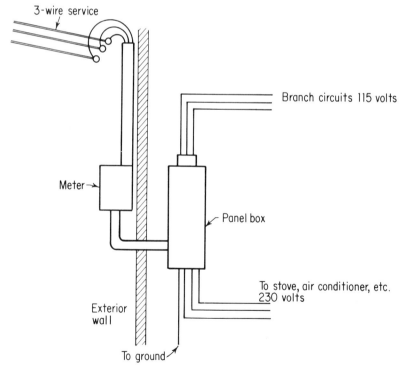

Small House Service.

THE ELECTRICAL INSTALLATION

electric range, or water heater, or any heavy motorized equipment. The average private dwelling now has 100-ampere service, which is double what it was just a few years ago. This service allows spare capacity for future devices or better lighting. Even in a fairly small house, the panel boxes should provide for at least 10 circuits. The initial extra cost for oversize facilities is very small and the architect should advise the owner that this is far cheaper than having to add to his wiring at a later date.

1B. THE SMALL BUSINESS BUILDING

In a small office building, provision must be made for the use of electric typewriters, calculators, and even small computers. Most office buildings now provide between 70 and 100 foot-candles of maintained light.

The architect engineer lays out his plan and in conjunction with the owner decides what facilities will be required for future occupants. He must make arrangements with the local power company to supply the necessary transformers and to run power lines into his building. (These should be underground if possible.) This may cost a little more but it can save a great deal of future trouble.

Transformers of various kinds cannot be picked off a shelf and therefore prior arrangements should be made with the power company. Such arrangements should not only include provisions for the availability of properly sized transformers but should also include the size and location of the transformer vault.[1] Usually the power company will provide a temporary transformer for the use of electricity during the process of construction; this is then replaced with a permanent installation when the building is enclosed and all internal wiring has been completed. The problem differs if such a building is being used by a large computer installation or for heavy use of electricity, but the principle is still the same. In the normal light-use office building (used for office purposes only), the current is brought from the power source through switches and smaller transformers to panel boxes. These boxes which contain fuses or circuit breakers, feed the circuit wiring for lighting and power. The circuit wiring can be brought into the floor area through the ceiling of the floor below, or through the floor itself if it is of corrugated steel of sufficient depth, or through certain types of partitions that have hollow bases suitable for pulling wire, or through underfloor ducts.

[1] A transformer vault is a fireproof space, very often underground, in which the voltage of the incoming street current is reduced, through transformers, to the voltage to be used in the building. For instance, the transformers may reduce 13,000 volts to 460 or 230 volts. The space is fireproof because often a transformer may "blow" through overload or just mechanical failure. Many transformers are filled with flammable insulating oil which could cause a serious fire.

THE ELECTRICAL INSTALLATION

Wall Mounted Raceways.

In the research building, medical building or laboratory, where there may be heavy use of electrical current at many places, the architect engineer must make provisions by means of underfloor raceways to bring current to wherever it is needed and for tapping such current at any reasonable location in an open floor area. He can also use bus-ducts which come in prefabricated lengths that may be run along a wall, and provide facilities for plugging in at almost any point. The underfloor raceway can either be incorporated in the floor system, as in the corrugated or cellular structural subfloor, or it can be separate steel, or concrete, or composition duct laid over a structural floor and covered with concrete. In any case this raceway can serve as the conduit for many pairs of wires. Each pair can serve an average of eight normal outlets (desk light, or four outlets used by electric typewriters, comptometers), so that raceways at reasonable intervals provide the utmost flexibility. Instead of the usual round wire, the bus-duct contains four copper or aluminum bus bars[2] that are simply rectangular conductors of electricity. The bus-duct comes in ventilated or completely covered units that can be bolted together to form any length. These bus-ducts can be plugged into at any point and carry from 225 to 5000 amperes, providing flexibility for the placement of all sorts of laboratory or research equipment.

In order to calculate the total electrical requirements, the architect engineer can use a table that is set forth in many codes, that gives the normal requirements of typical use structure in watts per square foot. To this he should add any out of the ordinary requirements and a factor for unknown future uses. In manufacturing buildings, where appearance is not of the utmost importance, wiring can be led through rigid conduit hung from the ceiling and brought down to the machine.

In general, the distribution of current in all such buildings follows the same pattern. It comes into the building at high voltage and is stepped down by means of a transformer to, let us say 277–480 volts. In this example, it would be distributed through the building at 480 volts and stepped down for lighting to 120–208 volts and specifically to 120 volts for small elec-

[2] 3-phase 4-wire.

THE ELECTRICAL INSTALLATION

trical office machinery or other equipment. It may be stepped down further for bells or buzzers. In any case, the basic theory behind all this stepping down of high voltage is that the higher the voltage the more efficient the transmission of electricity. It therefore follows that the reduction should occur as close to its ultimate utilization as possible.

1C. THE MULTISTORY BUSINESS OR RESIDENCE BUILDING

The calculation of the total electrical requirements of any major structure takes into account not only its presently planned use but the possible requirements for the future. For instance, in a large apartment building the plan may be for cooking done with gas. If later it is decided to change to electrical cooking, new electrical vertical feeders will be necessary as well as new circuits within the apartments. The same holds true for air conditioning. Some fine apartment projects, built quite recently, lack the electrical capacity to permit the use of window air conditioning units, leading to loss of tenants. The importance of keeping buildings contemporary electrically will be evident to the reader, and the architect engineer must anticipate future electrical needs.

1C1. *A Specification on Requirements*

In the office building, provision must be made for many possible uses that can be illustrated by the following list from a typical specification: The electrical work shall include but is not limited to the following: primary[3] and secondary voltage distribution facilities, including main switchboard, metering provisions, transformers, panelboards, motor control centers, conduit, cable, and bus-duct. Lighting and power systems including lighting fixtures and motor connections, underfloor power and telephone distribution system, telephone facilities including conduits and cabinets, elevator and escalator feeders, control wiring for remote control of mechanical equipment, smoke detecting system wiring, sprinkler alarm system wiring, emergency generator system, driveway lighting and landscape lighting. (Computer loading is a constantly increasing factor. This requires voltage stabilization as well.) In other specifications, the architect engineer, instead of listing the specific circuitry, might list the facilities requiring electricity, such as: the basement that will contain electric service rooms, telephone main frame room, pump room, fan and air conditioning equipment, building service areas, etc. The typical floors will be multitenant occupied. The thirty-third[4] level will contain the principal air conditioning equipment, heating plant, air handling fans, elevator machine rooms, etc.

[3] This is the voltage delivered to the building by the power company.
[4] This is a particular building.

THE ELECTRICAL INSTALLATION

The above specifications were written while the structure was being designed, the owner and architect engineer team having already determined the type of occupancy and the electrical requirements of the occupants. It had also been decided how the building was to be air conditioned, how air was to be distributed, how many elevators and escalators it was to have, and how it was to be lighted and heated. From these determinations it is now possible to calculate the total load and its distribution.

1C2. *Occupancy Requirements*

In the apartment building the planner, with the owner, must decide what the prospective tenant will require in the way of kitchen equipment, air conditioning, lighting, and laundry equipment. In luxury apartments there may be electrical broilers, several television sets, or motor driven exercise machines. Essentially, however, the requirements are not difficult to assess. The architect engineer should provide for a utility service with capacity for some expansion, and in addition provide an empty electrical conduit up through the building for future contingencies.

The large office building presents a more difficult problem. The electrification of office equipment is proceeding at a rapid pace. Large and small computers are being installed to overcome the shortage of clerical help. The requirements for lighting are being increased. The architect engineer can calculate his electrical load by a "watts per square foot" formula that will depend on the expected density of population, the foot-candles of light to be provided, and the kind of work to be performed. Special provision must be made for large computer installations, and for special air conditioning and air handling equipment to service such installations. Large drafting rooms or engineering offices require extra lighting. Advertising agencies may require heavy sound and closed TV installations as well as special lighting. Reproduction facilities or small printing installations use more than normal current. Obviously all this cannot be foreseen and this makes it doubly important to provide overcapacity in risers, and all main and branch circuits. Spare capacity must be left in underfloor ducts or raceways, in switchboards, in panel boxes, in the building service, and in vertical conduits.

1C3. *Equipment Requirements*

The planner then adds all the building equipment load to the electric capacity required by the tenant. Here he is on fairly firm ground. The sizes of all his motors for fans, pumps, elevators, and heaters are known. At some future date such equipment may be changed to require more power but this is not likely, and in any case the equipment is large and concentrated in easily accessible areas. The lighting may have to be increased and spare capacity should be provided. The power company should be approached

THE ELECTRICAL INSTALLATION

promptly when the needs of the occupants and the requirements of the building equipment have been determined. It may be necessary for the power company to run new feeder lines to the location and new transformers require considerable lead time. The author knows of instances where permanent power was not available when required, causing a delay in occupancy. Equipment such as elevator motors or compressor motors require a permanent power source.

1C4. *The Distribution*

The distribution system of various types of small buildings has been previously described. The system for an apartment building is relatively simple, and does not require much flexibility. The size of the apartments is fixed and provisions made for any extra electrical equipment tenants may later use. The building-owned equipment such as stove, refrigerators, or air conditioning units is fixed as is the large equipment such as pumps, boilers, and fans. The building service usually enters underground into a transformer that converts high-voltage street current into low-voltage building current (120/208 for instance), and the electricity is conducted up to the various floors through wire cables enclosed in conduits. Depending upon the size of the building, the current from these cables is taken off at one or more locations on each floor where it terminates in a panel box. This box then distributes the current into circuits that may feed a number of apartments. The circuiting is so arranged that air conditioners or stoves are on 208 volts while lighting outlets and lighter kitchen units are on 120 volts. This simple circuit distribution is of recent origin and can be done this way because under the new regulations in some localities, the tenant can have electricity included in his rent. Another compelling reason is that the cost of labor and material in the electrical industry has risen so rapidly that the cost of any excessive amount of wiring is prohibitive. In localities where the tenant may not have electricity included in his rent, the wiring is much more complicated because each apartment must have its own separate circuit, its own fuse or circuit breaker box, and its own electric meter. If such meters have to be located in the basement of the building where they can easily be read, the extent of the wiring can be readily imagined.

The electrical system of a major high-rise or large office building must have complete flexibility and heavy built-in extra capacity or provisions for its installation. The tenant is constantly rearranging his space. He is consolidating some departments and moving others. In a growing organization it is normal to change 25% of the space in one year. This constant flux demands main feeders and branch circuits to be of adequate size and capacity to meet maximum future needs. Let us see how this is done. We will take a typical building in New York City as an example, although all such buildings are alike except for the voltages or the peculiarities of the local code. The

THE ELECTRICAL INSTALLATION

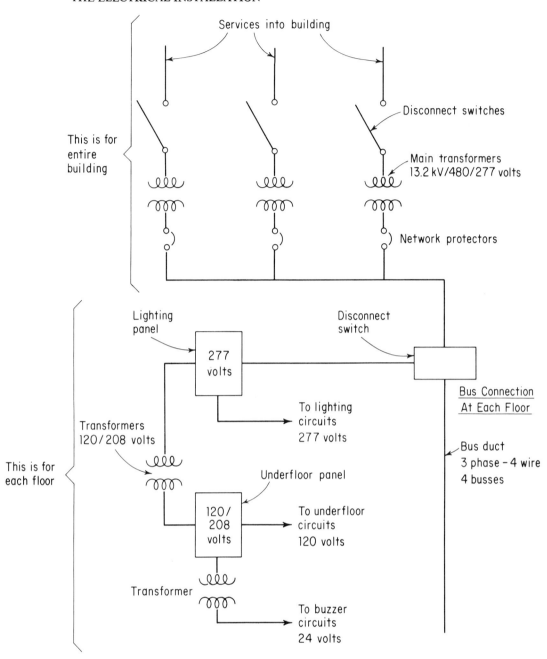

Typical Large Business Building Distribution System.

current enters a transformer vault that is usually located under a city sidewalk and is the property of the power company. The street service is carried through mains at 13,200 volts, which is stepped down within the transformer vault to 277/480 volts. This is a three-phase four-wire service that supplies 480 volts among any of the three current-carrying wires and 277 volts between them and the ground wire. This current is then carried through service switches to a main switchboard. From the switchboard the current is borne in either horizontally suspended enclosed bus bars or large cables running through pipes (called *conduits*), and vertical enclosed bus bars that carry the current up through the building in a shaft, the shaft going through an "electric closet" on each floor. The electrical load is then distributed as follows: On each floor there are taps into the bus duct or cables and other cables carry the current from these taps into a panel box, which through circuit breakers, or fused switches, distributes the current to numerous lighting circuits at 277 volts (which is the voltage between any one live bus and the neutral bus.) From this lighting panel the current is stepped down by a transformer to 120/208 volts and then through another panel box it is sent to underfloor circuits for use by business machines. The voltage can be stepped down further for buzzers, bells, or sound systems. The highest practicable voltage is carried as far as possible before it is stepped down, to attain top efficiency.

The distribution system to the building equipment is run through either a separate set of bus-ducts used for power only or through vertical risers carrying heavy cables. All such equipment runs at 480 volts and taps can be taken off the main feeders without going through a transformer.

2. TYPICAL POWER COMPANY DISTRIBUTION

At this point it might be useful for the present or prospective architect engineer to acquire a general idea of how electrical power is distributed. First, different localities distribute power at different voltages and these are constantly changing as population density increases, as more electrical equipment is used, and as the local power company modernizes its equipment. Until very recently in the author's small suburban hometown, power was sent to a centrally located transformer station at 13,200 volts. It was stepped down to 4160 volts and sent out along overhead lines to feed the community. It was stepped down again to 115/230 volts by means of transformers located on street poles and fed into the private homes, stores, or small office buildings. Because of the long distances from this single power source and the comparatively low voltage, there was a considerable variation in this voltage and lights flickered or burned out, sometimes the TV screen seemed to get smaller, and if an ice storm caused the line to break, it was shattering to see your next door neighbor with lights and heat while you were in the dark and cold because you were on the wrong side of the break. This has all changed. Now the overhead lines are fed from two directions and they are run at 13,200 volts. The result is steady voltage and reliable supply.

In urban areas where there is density of population and transportation, and where the business community cannot function without electricity, the power companies have taken extraordinary measures to assure a reliable

THE ELECTRICAL INSTALLATION

supply. The great blackout of the Northeast of November 1965 cannot occur again. For instance, in New York City the many generating stations serving the central business area have been combined in a loop network with 50% overcapacity in underground cables and building transformers, and the connection between outside power sources and the city network can be automatically disconnected within a fraction of a second. Current is distributed at 13,200 volts into the various buildings through conveniently located central transformer stations. Because of the many operations that require constant voltage (a computer can be thrown off very badly by a serious voltage fluctuation), the power companies take many precautions to insure continuous flow, but transient fluctuations do occur. If it is imperative for the architect engineer to furnish constant voltage, power companies recommend motor generator sets[5] to buffer out fluctuations.

This section can be concluded with the statement that utilities everywhere are constantly improving their service and reliability and the architect engineer must be constantly aware of such improvements and changes in service so that he may take advantage of them in the design of the electrical installation.

3. SELECTION OF EQUIPMENT

The specifications for any electrical system usually give the electrical contractor a choice of several manufacturers for each piece of equipment; for the wiring, the lighting fixtures, the conduit, the panel boxes, and so on. If the specification is well written, it confines the choice to reliable manufacturers and it should always ask the contractor to furnish a list of equipment for prior approval by the architect engineer. For instance, a typical specification for wire carrying 120 volts may state that it must have 600-volt National Electric Code insulation, be color coded, and come in unbroken containers. The specification for a panelboard cabinet will mention the gauge of the metal, the overlap of the door over the opening and how the cabinet is to be mounted. The type and materials of circuit breakers and switchboards are specifically described and tied to national and local codes and trade association standards such as the National Electrical Manufacturers Association. This is all done to protect the owner against substandard material and methods. The architect engineer should be aware of the latest products of manufacturers of electrical equipment. The sales engineers of such companies are more than willing to demonstrate their products and catalogs, and brochures are always available. The architect engineer must be aware of the entire range of contemporary products.

[5]Solid-state inverter equipment is gradually replacing the M-G type.

4. LIGHTING

The lighting of a building is so important that it must be carefully considered as a subject in itself. The demand of business for good lighting without eye-strain or other detrimental physical effects has received the attention of many experts in the fields of lighting, medical, and environmental research.

The first step in providing proper lighting is to determine the use of the structure. A conventional business building requires lighting that is nonglaring and is evenly distributed over the working area, which is usually considered at desk level. In this connection, manufacturers of business furniture emphasize the importance of the material of working surfaces. They make such surfaces non reflective and of a color that is not obtrusive but pleasing to the eye. The author knows that if he suggests an acceptable lighting level for ordinary business use, he will be subject to arguments from all companies and associations whose business is lighting. Nevertheless the lighting level of a number of very handsome and successful recently constructed highrise office buildings, with which the author has had some connection, can be mentioned. The buildings have a lighting level of from 65 to 85 foot-candles maintained[6] at desk level. There have been no complaints about eye strain, glare, or insufficient light from the thousands of employees and dozens of enterprises occupying these buildings. Of course, if the occupants are to be engaged in drafting or design work, or research, or other activities (color selection, artwork, textile design, etc.), the lighting intensity should be higher (up to 150 or more foot-candles) and in fact the entire lighting system should be designed to fit these requirements.

This section, however, will confine itself to conventional requirements. The architect engineer and the owner having determined the required lighting level, must then decide how it is to be attained. They must consider the ceiling height, the spacing of the fixtures (which should be a product of the modular concept of the building), and the size[7] of the spaces to be illuminated. When the above determinations have been made, the proper size of fixture and the number and wattage of the fluorescent lamps to produce the necessary output of light is determined. There are all sorts of tables that give this information, once you know what you want.

The construction of the lighting fixture itself and its lens must be carefully specified. It must be sturdy; the tubes must be easily accessible; the lenses must be sturdy, and of a construction that will evenly distribute

---

[6]Maintained means the foot-candles produced by a lighting system toward the end of the useful life of the lamps. The initial higher intensity is burned off after about 100 hours.

[7]Due to "light spill" from one source of light to another, a large space will produce more light per fixture than a smaller space that may encompass any number of fixtures from one up to eight or ten. (This is an arbitrary figure.)

the light, and of a material[8] that will not discolor, and the fixture must be deep enough so that the tubes themselves will not be easily discernible through the lenses (hot spots).

A specification for fixture construction would contain the following: Fixtures shall be of metal of not less than 20-gauge[9]. They shall have reinforcing members for rigidity. All metal parts shall be spot welded or joined with sheet metal screws and be free from "light leaks." They shall come equipped with reflectors, sockets, ballasts, and wire channels and shall be fitted with 6 feet of flexible conduit with fixture wire drawn in with 1 foot of slack. Wireways for rows of fixtures shall be underwriter approved. Plastic lenses shall be of the prismatic acrylic type. Lenses shall be of 100% virgin acrylic plastic meeting ASTM specifications and shall be one piece fully injection[10] molded. A good specification should also call for a fixture demonstration test to be conducted in a facsimile office space.

Besides acting as sources of light, fixtures now serve also as distributors for air in which case there are slots on their sides which either send conditioned air into the office space or exhaust it. In an "all electric" building, to be described under "Air Conditioning," the fixtures are used to circulate water that uses the heat of light for warming purposes.

The lighting fixture should, as far as possible, be maintenance free and of the best possible design to produce the very important, required light intensity and equality of distribution.

This section has confined itself to general illumination produced by fluorescent lighting. Of course, the incandescent light is a very important source of light, not only in the private residence but in business or social activities where its warm light and decorative qualities can be fully exploited. Banks, restaurants, building lobbies, places of worship, auditoriums and many other structures take advantage of its qualities.

One last word about the quality of light. Even general fluorescent lighting can, by the use of warm white or cool white lamps, make people look good; and constant improvements are being made. People who look good to themselves and others, perform better work.

5. EMERGENCY GENERATORS

It was stated previously in this chapter that the great Northeast blackout of 1965 can never occur again. This is true enough but occasional small

---

[8]Polystyrene plastic is apt to discolor; acrylic plastic costs more but keeps its clarity.

[9]This gauge is in some big city codes; others do not specify gauge but it is recommended that serious consideration be given to the use of this gauge.

[10]Plastic lenses can be produced by extrusion in which the plastic is squeezed through rollers, by ordinary molding, or by injection molding by which plastic is injected into a mold by pressure. This is supposed to produce the sharpest lens prisms and is the most expensive.

THE ELECTRICAL INSTALLATION

blackouts do occur in spite of every precaution. Nobody can prevent a transformer from blowing suddenly or a laborer from running his pneumatic drill into a power feeder. Consequently, prodded by the blackout, many cities have amended their codes and require new buildings of certain sizes and heights to install emergency generators. Such generators must have sufficient capacity to activate emergency stair and corridor lighting, fire alarm and sprinkler alarm systems, a sound system, a fire pump to provide water for fire fighting, and at least one elevator that stops at every floor of a high-rise building. Although no code has forced older buildings to do this, many wise owners are installing such generators. They can be driven by diesels or gas turbines and can start immediately when the major power source fails.

The author would like to reemphasize the constant and dynamic changes in the use of electricity. More and more of the clerical, manufacturing, and medical processes are using electricity and the forward looking planner will be aware of this and prepare for it.

CONTENTS

1. HEATING
   A. The Private Residence
      1. Determination of the Requirements
      2. Standard Heating Systems and Variations
      3. A Typical Specification
   B. The Small Multiple Dwelling or Business Building
      1. The Requirements
      2. Choice of the Plant
      3. The Distribution
   C. The Large Office Building
      1. The requirements
      2. Choice of the plant
      3. The distribution
   D. Choice of Fuel
   E. Building Code and other Requirements

2. VENTILATION AND AIR CONDITIONING
   A. Required Ventilation
   B. Determination of Cooling Requirements
   C. Cooling Plants
      1. Centrifugal compressor systems
         a. Specifications and description of the cycle
         b. Equipment sizes
      2. Absorption liquid chillers
      3. Reciprocating units
   D. The Air Distribution System
      1. Zoning
      2. Discussion of proper location for the plant
         a. The chiller units and heating units
         b. The fan rooms
      3. The high and low pressure system
      4. Single duct system
      5. Double duct system
      6. Perimeter window units
      7. Exhaust system
      8. Sound abatement
   E. The Control System
   F. Water Treatment and Filters

3. THE "ALL ELECTRIC BUILDING"—A VARIANT

9

HEATING, VENTILATION, AND AIR CONDITIONING

1. HEATING

In the north temperate zone in which this country is located, almost every shelter, whether it be home, factory, or place of business, has to be heated at some time of the year. Many structures have to be ventilated artificially as well, and nearly all structures used for business are now cooled in the summer.

How the structure is to be heated and what fuel is to be used depend upon the climate, the size of the structure, and the availability and comparative cost of various fuels. In order to calculate the size and kind of heating plant required, the architect engineer must consider several things. First is the use to which the structure is to be put. A residence should be warmer than a gymnasium, and an office warmer than a plant where workers move around. Having established the inside temperature he wishes to maintain, the planner then determines from Weather Bureau records the range of the outside temperature. For instance, he should plan a 70° temperature for 0° outside for a residence or office building in the North Atlantic States. It is true that there are very few 0° days, but a slightly oversize plant does not cost much more to install and is more economical to operate. When the temperature difference has been established, the next step is to calculate the amount of heat necessary to maintain this temperature. Consideration must be given to the type of exterior materials. Various materials transmit heat at different rates. The presence or lack of insulation and the types and construction of openings such as doors and windows must be considered.

HEATING, VENTILATION, AND AIR CONDITIONING

1A. The Private Residence

1A1. *Determination of the Requirements*

Let's take a private residence as an example: First the desired temperature difference is established; next the material of the exterior walls and roof are determined. Each room is then examined separately to determine the heat to be put into it. These rooms are added to all the other spaces for the total requirement. Handbooks give tables on the number of BTU[1] per hour per square foot of area which every kind of material will transmit. The planner should also ascertain leakage of the outside air through the various kinds of openings at different wind velocities. The use of each room must be considered; for example, a bathroom should be warmer than a kitchen. The exposure of each room must be taken into account. In our climate the north side requires more heat. With all these factors determined, the planner can determine how many BTU per hour he must put into the space to keep it at $X°$ when it is $Y°$ outside. By adding all the room requirements and some percentages for loss in transmission from the central heating plant (if there is one) to each space, the architect engineer determines his total requirements and the size of his plant. He now comes to the determination of the kind of plant and the fuel to use.

1A2. *Standard Heating Systems and Variations*

In northern climates a central heating plant is almost always used. In a smaller residence this could be either a hot water or steam boiler, or a forced warm air plant. A hot water system that circulates hot water through either cast iron radiators or base board radiators[2] is capable of maintaining an even temperature because the source of heat (the water) is kept at a comparatively low temperature (200° or less) and the radiating surfaces are large. The radiators are designed to transmit heat by radiation and the convection of air over and through their warm surfaces. Because of the large warm metal surfaces the heat in a room is not lost quickly.

Another advantage of a hot water installation is that it can be zoned so that various parts of the house can be warmed to different temperatures. This is done by installing a thermostat in each zone. The thermostat activates a circulating pump that works only when heat is called for. If one zone calls for more heat than another, the zone pump will send more hot water to it.

---

[1] The British Thermal Unit (BTU) is a standard of heat measurement. It refers to the amount of heat necessary to increase the temperature of one pound of water, one degree Fahrenheit.

[2] Baseboard radiators are heating pipes to which thin copper, aluminum or steel fins are attached. The hot pipe heats the fins which through their large area, heat and circulate air through convection currents.

HEATING, VENTILATION, AND AIR CONDITIONING

Hot water can also be circulated through coils of piping laid under a concrete floor slab or over a ceiling. This type of heating is called *radiant heating* and although it has been very popular, it is not recommended. Piping does leak and it is extremely difficult to discover the location of the leak and prohibitive to repair when it is under a concrete floor slab. Radiant heating is still being installed in development housing because of its economy. The coils of pipe can be fabricated in large units and these units are laid in place and connected on the site with the use of very little skilled labor. The builder then pours a concrete slab over the pipe and his heating system is complete except for connecting a heating boiler.

The steam boiler, which is now very rarely used in private homes, raises the water temperature to above the boiling point and sends the steam through riser pipes to cast iron radiators. The steam condenses as it gives up its heat to the radiator. The condensate water then flows back through inclined pipes to the boiler. There are variations such as a two-pipe system that uses a vacuum return pump to suck steam through the radiators and return it to the boiler.

The forced warm air furnace is a heating unit surrounded by a shell through which a fan circulates air that is then sent through sheet metal ducts to the various rooms. The air is returned to the furnace through one or more centrally located ducts so that it is constantly circulated. The partly warmed return air is reheated and sent back. Fresh air is supplied by natural leakage of outside air which takes place in any normally built house. The warm air can be humidified in the winter and a warm air system is capable of being converted to air cooling in the summer. Unless the system is well balanced the heating is not as even as with hot water, but many fine homes use either system.

There are many climates where central plants are not necessary. If the heating load is light and the heating season does not last long it is practical to use floor furnaces, which are self-contained heating units installed under the floors of the various rooms. If the rate for electricity is low and the climate is relatively mild, it is also possible to use electrically heated wall units.

It should be kept in mind that electrical resistance heating, where electrical energy is turned directly into heat, is the most expensive way of using electricity.

There are other methods for heating the small structure, some of them highly experimental. The use of solar heat is always fascinating to the architect engineer and owner. It represents something for nothing. The trick is to use it economically. There have been examples of large tanks of brine or other solutions placed on or just under a roof where they will get the full benefit of the winter sun. The warm solution is circulated through large radiating surfaces. Unfortunately, the cost and weight of the tank and the radiators and the fact that sunshine is undependable has made such installations unsuccessful. Another apparently free source of heat is the heat pump.

HEATING, VENTILATION, AND AIR CONDITIONING

This unit works on the principle of any refrigeration unit (this will be described later). The cycle of the pump can be reversed so that it extracts heat from the interior in the summer (thereby cooling it), and extracts heat from the outside air in the winter and transfers this heat to the interior. Before considering such an installation, the planner must take into account the cost of electricity, the cost of the installation, and the existent climate. In severe climates the winter heating requires more energy than summer cooling and therefore the unit must be oversized unless a supplementary heat source is furnished. It must also be remembered that for a six-room house, a 2-ton electrically-driven compressor will be required to run many hours a day. The heat pump does its most economical work in a mild climate where the heating and cooling loads are nearly equal and electricity is cheap. The heat pump should also be considered in situations where the internal heat gain from people and equipment is almost enough to overcome low outside temperatures.

1A3. *A Typical Specification*

A typical specification for the heating of a single-family dwelling would call for the following items: To install a heating system capable of maintaining an interior temperature of X° at Y° outside. The heating system to be a one-pipe forced hot water system, or a forced warm air system or whatever system is decided upon. The heating plant to be of sufficient capacity plus X% to carry the entire load and to, if specified, be capable of carrying the hot water load; it shall be of a certain manufacture (several choices are given). The burner to be of specified capacity and of manufacture as specified. The safety devices and controls are specified. The circulating pumps or fans are specified. The sizes of the ducts or radiators and their manufacture are specified. If it is a comparatively small house (seven or eight rooms or less), a good deal of the calculation of capacities, pipe sizes, duct sizes, etc. can be left to the local plumbing and heating contractor. He will take the plans and specifications to his supplier who will get the manufacturer of the specified equipment to do the calculations. It is well, however, for the architect to be able to check these figures. In larger homes the architect engineer should supply drawings and schedules of heating loads but the specification should still state that the contractor must furnish and install an adequate system and thereby make him responsible for checking the figures.

1B. THE SMALL MULTIPLE DWELLING OR BUSINESS BUILDING

1B1. *The Requirements*

The calculation of the heat load for a small building or multiple dwelling is essentially the same as for a residence. The ratio of the area of exposed walls to enclosed floor space is lower for larger structures so that the

HEATING, VENTILATION, AND AIR CONDITIONING

heat loss is less. The number of people occupying the space and their activity must be considered in buildings other than dwellings. It is interesting to note that a normal adult performing office work exudes about 400 BTU per hour. This is higher when the occupant is engaged in strenuous activity. That is why auditoriums with many bodies close together, gymnasiums, factories or laboratories need less heat than dwellings. A medical building in which patients are examined needs more heat. The architect engineer can calculate his heating load from the combination of all these factors, and then choose the central plant that will satisfy these requirements, and use the most economical available fuel.

1B2. *Choice of the Plant*

The choice of the heating plant for a small building depends to some extent on whether the building is also to be air conditioned. Generally, an oil- or gas-fired hot water or low-pressure steam boiler is used. This is usually a packaged unit manufactured by a half dozen reliable companies and the proper size can be virtually picked from a catalogue. The specification for such a boiler is described in Section 1C2 of this chapter. There are packaged units available for both heating and cooling. These units are for small installations and are generally not available over a 25-ton capacity for cooling. At approximately 350 square feet for a ton of cooling, a 25-ton unit can cool a maximum of 10,000 square feet.

1B3. *The Distribution*

Heat in a small building that is centrally air conditioned can be distributed by air that is heated at a central point by hot water or steam, and then sent through ducts to the interior and perimeter of the building. If the building is

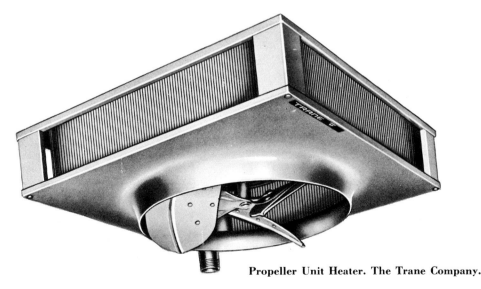

Propeller Unit Heater. The Trane Company.

HEATING, VENTILATION, AND AIR CONDITIONING

air conditioned by window units, the hot water is usually supplied to under-window or baseboard radiators through a system of riser pipes. Circulating pumps keep the water flowing and the flow is controlled by thermostats. The greater the call for heat the wider the control valve opening. Large areas such as stores or factories are often heated by unit heaters which consist of a heating coil through which air is blown by a fan mounted in the same casing. Such unit heaters are capable of heating as much as 5000 square feet of space, depending on the height at which they are mounted, the temperature of the water or steam, the type of building in which they are used, and the kind of work being performed in the building. For instance, in a garage or other building where large doors may be opened frequently, it is well to furnish a unit heater that blows directly down on the opening and provides a curtain of air. The capacities of unit heaters are determined by their BTU hour output. It will be well for the architect engineer to remember that workers nowadays consider comfortable working conditions as one of their prerogatives.

1C. THE LARGE OFFICE BUILDING

1C1. *The Requirements*

The heating requirements of the large office building present an interesting and fairly intricate problem to the architect engineer. Without exception the heating design is tied in with the ventilation and cooling system. The calculation of the heating load starts with the same basic premises as it would for any structure—the loss of heat through the exterior walls, the leakage of air through the openings, the population load, and its normal activity. In addition, in constructing a tall building, the effect of the wind and sun must be carefully considered, as well as the heat losses in the extensive distribution system. Because the heating is associated with the ventilation and cooling, the heat is always transmitted by means of air that is warmed by passing over warm surfaces, and blown into the various areas. The air is then recirculated, but because of the population load and the tight construction of the present-day building, fresh air must be added. Each one of the thousands of heating surfaces must be calculated and after adding for heat losses in distribution, the architect engineer can determine the total heating capacity.

1C2. *Choice of the Plant*

The heating plant to be chosen for a large structure can range from a simple low-pressure hot water or steam boiler to a heat exchanger, or a high-pressure steam plant that supplies steam for turbine-driven compressors for air conditioning, and uses the "left over" low-pressure steam for heating. The architect engineer must make a very careful calculation of the cost of the installation of the various plants and of the relative cost of the electrical energy available. If the cost of electrical energy makes it more economical to use electric motors for running the large air conditioning compressors, then it follows that the hot water or low-pressure steam for heat will be supplied by a low-pressure boiler. Such boilers are usually of the "packaged" type, which means that they come to the site ready to install. A typical specification for such a boiler would be as follows: Each boiler shall be a multipass horizontal fire tube boiler of the specified capacity. It shall be pre-assembled and fire tested at the factory using the specified fuels. It shall be ready for immediate mounting and ready for attachment of water, fuel, blowdown, electrical and vent connections. The boiler construction shall conform to standards set by the American Society of Mechanical Engineers (ASME) and shall be designed for 15 pounds pressure. The specification describes the safety devices such as stack switches (which turn the burner off when the temperature exceeds a certain point), safety valves, which blow when the pressure exceeds 15 pounds, and boiler trim such as gauge glasses, pressure gauge, and so forth.

In projects where energy produced by electricity is not as economical as energy produced by certain fuels, the architect engineer considers the installation of a building power plant that produces high-pressure steam to run the air conditioning compressors and to heat the building. He must consider the capital expenditure that must be made for such boilers, the steam turbines, the extra size of the boiler room, and the required chimney or stack. He must also remember that all codes require a qualified licensed watch engineer to always be on duty when a high-pressure boiler is in service. The high-pressure boiler specification closely follows the specification for the low-

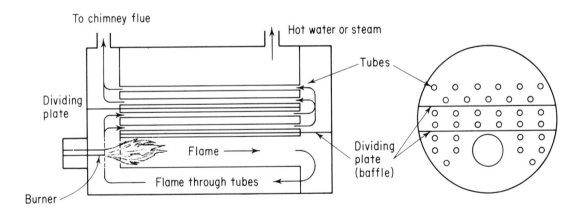

The Flame Is Sent Through A Large Tube Then Sent Back
And Forth Through Small Tubes By Means Of Dividing
Baffle Plates. All The Tubes Are Surrounded By Water

Fire Tube Boiler.

HEATING, VENTILATION, AND AIR CONDITIONING

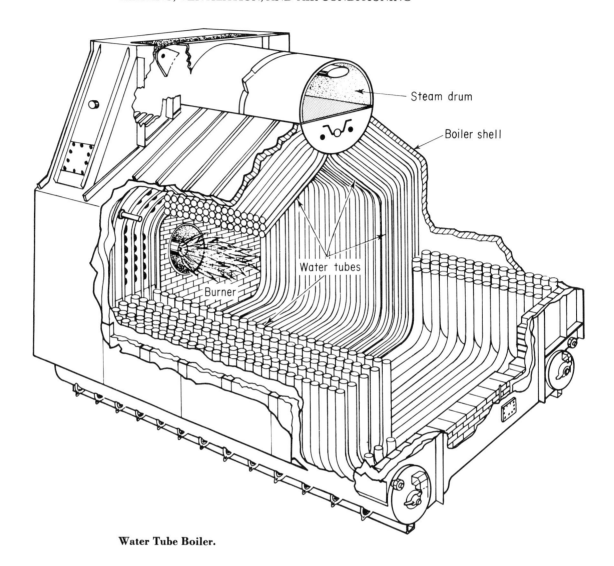

Water Tube Boiler.

pressure one. The construction specification is more detailed. The boiler must meet A. S. M. E., National Underwriters and local code standards. It calls for a multiplicity of safety devices. Boiler inspection and testing must take place more often. The boiler is usually "packaged" and pretested and is of the water tube, instead of the fire tube type used in low pressures.

In a number of cities high-pressure steam is available in central areas. This steam is produced by a public utility and distributed through mains under the public streets. When such steam is available it can save the installation cost of any boiler plant, chimney stacks, and a great deal of space. Such steam is used for driving turbines for the air conditioning compressors. The steam that is exhausted from the turbines is run through coils in a heat exchanger where the steam heats water which is then pumped through the building. In winter the steam runs through a pressure reducing valve directly

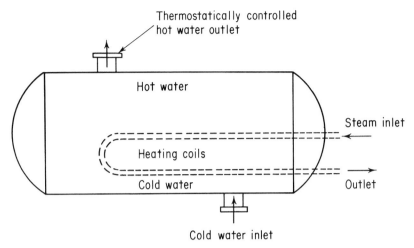

Heat Exchanger.

to the heat exchanger. There is one other way of heating a building. In an "all electric" building, the winter heating and summer tempering can be supplied by a combination of heat pumps, heat exchangers, and the heat of light. This will be described later in this chapter.

1C3. *The Distribution*

The manner of distributing the heat to the various portions of the building depends on the total design of the air or weather conditioning. In some designs the hot water is sent to a series of window units around the perimeter of the building. These units are built to force air over the resultant warm surfaces. In other cases the hot water is supplied to large coils located at central fan systems which again blow air over the warm surface and then send the air through ducts to the perimeter areas. In all cases the interior areas are warmed where necessary by air that has been warmed at a centrally located heating unit. Hot water is also distributed to tempering or reheat units that are used in summer cooling and which will be described in Section 2C1a under air conditioning. The low-pressure boiler can also be a steam boiler, in which case the steam is used to heat water for the perimeter window units and the steam itself is used for the tempering coils. While there may be some efficiencies in the distribution of steam over hot water, this system is not generally used.

1D. CHOICE OF FUEL

The choice of the fuel is entirely dependent on the economics of the installation. Fuel represents energy to be used for heating, for driving motors or turbines, and for lighting. Whether the fuel is used directly in the structure or at the public utility plant to generate electricity, or to furnish high pressure steam, its total cost and the cost of the proper installation are the primary considerations. In some areas where natural gas is plentiful and reasonably priced, it is used to heat small homes by means of underfloor furnaces. Gas is used to drive steam turbine driven refrigeration compressors by means of heating water to high-pressure steam or by means of high-pressure gas turbines. Gas heats water directly or by exhaust steam from such turbines.

In many areas, and especially in those located on navigable waters where fuel oil can be delivered by tanker or barge, such fuel oil is used extensively for heating small homes and in its heavier grades is used for low- or high-pressure boilers of large size. There are installations where natural gas is used in the summer "off peak" for making high-pressure steam for air conditioning and fuel oil is used for heating in the winter. In some sections coal still is used extensively. Many large power plants use pulverized coal in tremendous boilers that generate steam at 1000 psi to drive turbines and generate electricity. The use of electricity as a direct heating medium by the use of wall panels is still considered an expensive way to heat in most communities. What the architect engineer and owner are buying is energy, whether it be street steam, fuel oil, natural or manufactured gas or oil, and the comparative cost per unit must be known before a final decision is made.

1E. Building Code and Other Requirements

The planner must always be aware that the heating plant is subject to code restrictions. Even in private homes in small towns the code prescribes chimney or flue sizes for various sizes of boiler installations, the allowed sizes and locations of fuel storage, the fire resistant qualities of the immediate surroundings, and it specifies that all burner installations must be of approved manufacture and must be inspected. As the boilers get larger and the pressures higher, the codes become more restrictive. They require low water cut-offs, water gauges, pressure relief valves, stack switches, and other safety devices. The manufacturer's or the ASME requirements for their construction also become more stringent. All codes call for periodic inspection of high-pressure boilers by the authorities, and also call for a duly licensed attendant at all times that these boilers are being used.

2. VENTILATION AND AIR CONDITIONING

There are many situations where artificial ventilation is necessary for reasons of health, or to provide a suitable environment for the activity taking place. Where structures are air conditioned, that is, where air is used

HEATING, VENTILATION, AND AIR CONDITIONING

as the medium of transmission for cooling in the summer and warming in the winter, then conditioning and ventilation are so interdependent that they must be considered together.

2A. REQUIRED VENTILATION

The most elementary form of obtaining fresh air is obviously by opening a window or other opening to the outside. It therefore becomes a matter of concern to the public authorities that there be sufficient openings in any structure to the outside air, in order to maintain the health and welfare of the inhabitants. Such requirements take into account the sizes of the areas to be ventilated. In structures other than dwelling units where there are large interior areas where numbers of people may congregate, the codes are very explicit about mechanical ventilation. The basic requirement is always the amount of air that must be supplied and exhausted and is expressed in cubic feet per minute (CFM). The tables of requirements list all sorts of uses and the requirements for each based on area and population. One of the newest codes, that of New York City, expresses the requirements for such air by using a factor composed of population, area and volume of the space, and the openings to the outside air.

Mechanical ventilation in its simplest form is maintained by motor driven fans that force air through sheet metal ducts which are the correct size to furnish air where it is needed. The manufacturers of such fans have tables to show how many CFM a fan will produce at a given speed. It would seem simple to obtain the proper fan sizes. However, duct sizes, the velocity of air through them, and losses through friction complicate this. If only ventilation is supplied the air must be warmed in cold weather. The architect engineer must also take into account sound insulation which will be discussed later.

2B. DETERMINATION OF COOLING REQUIREMENTS

The air conditioning of a building can be better described as climate manufacturing. The inhabitant of any structure, whether dwelling or place of work, wishes to be comfortable all year round. In the winter he wants to be warm and have the air slightly humidified, and in the summer he wants the air cooled and dehumidified. Not so long ago one could keep warm by the use of hot air furnaces or steam radiators, and cool by means of opening a window and turning on a fan. With the advent of the room air conditioner or small central plant for the dwelling, and the major central refrigeration and ventilating plants in large structures, all year comfort can be obtained. The competitiveness of present-day business, its requirements for skilled employees, and its demands on such employees requires an environment that furnishes complete bodily comfort.

The architect engineer, in calculating the sizes of cooling plants and the capacity requirements for fans, ducts and cooling and heating coils, takes the same factors into account as he does for heating. He calculates the warming effect of the outside air through the building walls or openings; he takes account of the population and its activity; the lighting load and any machine load; he makes allowances for various exposures, sun load, and the effects of wind. A cooling plant is designed to meet certain standards of temperature, humidity, and air supply. These standards have been arrived at through experimentation and experience and are meant to provide the most comfortable interior weather to people engaging in normal activities. Engineering specifications generally do not mention these standards, because the capacities and sizes of all equipment that is specified are calculated to produce this norm. However, for the general public, owner, or tenant, the comfort norm is expressed as follows: to provide an indoor temperature of 70° and 50% relative humidity when outside temperature does not exceed 95° dry bulb and 75°F wet bulb. Please note that these standards are expressed as temperature and humidity differences rather than in absolute figures. For normal use, to design to an absolute figure such as 72° and 50% humidity inside, at any outside temperature or humidity, would call for an absurdly oversized plant and would not be good for people going back and forth between such extremes. One other design norm is the determination of the "dew point," which is the temperature at which the moisture in the air condenses. The dew point can be determined from the same table that gives the relative humidity. The designer must be very careful to see that the interior humidity does not rise or the temperature drop to the point where moisture condenses on windows or walls.

Of course there are cases where an absolute temperature and humidity must be maintained. Computer mechanisms, processes where small electronic parts are assembled, and certain laboratory or research projects require this. If these activities take place within a larger structure in which normal activities also occur, the architect engineer will carefully heat-insulate these areas and furnish them with their own system. Now, because all the heating and cooling is done by means of air, he must get the air to the right places in the right quantities and get rid of the stale air. He must allow for loss of efficiency if his ducts are too long. If they are too small they will whistle because the air goes through at too high a velocity, and if they are too large he is wasting money.

[3]The relative humidity of air can be measured by a sling psychrometer which consists of two thermometers mounted side by side, one of which has a wick surrounding its bulb. The wick is saturated and the two thermometers are whirled rapidly. The evaporation on the wet wick produces a lower temperature than will be read on the dry thermometer. The two temperatures indicate the relative humidity, which can be read from a psychrometric chart.

HEATING, VENTILATION, AND AIR CONDITIONING

When the architect engineer is all through with his calculations he will know the size of the cooling plant in tons of refrigeration,[4] the size of the fans, and the sizes of the main distributing ducts and the branch ducts. Depending on the type of air distributing system to be used, he will know the sizes of the window units or mixing boxes or ceiling slots. With all these required capacities determined, he is ready to choose the central cooling plant and the air distribution system.

2C. Cooling Plants

Cooling plants come in many varieties and capacities from the window air conditioner to the multi-ton central plant. However, they all work on the same refrigeration cycle which can be described simply in a window unit. The window air conditioning unit contains a motor driven compressor that compresses a refrigerant gas, such as Freon. The compressed gas is cooled by a fan and the cooled high-pressure gas, which has turned to high-pressure liquid is sent through an expansion valve and becomes low-pressure liquid. As the cooled low-pressure liquid vaporizes, it absorbs the heat of vaporization from the air passing over the coils in which the vaporization is occurring. In passing over these cold coils, the air temperature has also dropped below its dew point so that it has dropped part of its moisture in the form of condensate water on these coils. The air that is blown into the room is therefore not only cool but dehumidified.

2C1. *Centrifugal Compressor Systems*

Although reciprocating compressors are used extensively in small plants, the most commonly used machine for central air conditioning is the centrifugal compressor. The heart of this machine, and its only moving part, is the compressor, which is driven by an electric motor or a steam or gas turbine. The compressor, chiller and driving motor usually come packaged as a unit in smaller installations (from approximately 50 up to 600 tons), and come separately and must be field connected in the larger installations.

2C1a. *Specifications and description of the cycle*

A typical specification for a compressor unit would state the following: Compressor shall be hermetic, two-state, direct driven, operating at no more than 3600 rpm with a rotor that shall be statically and dynamically

---

[4] A ton of refrigeration is the amount of heat extraction necessary to freeze one ton of water at 32° to one ton of ice in 24 hours. This is expressed in BTU and means the number of BTU's or amount of heat (288,000 BTU per ton) that will be extracted from the given area every 24 hours.

balanced; chiller shall be of shell and tube type with water side tested for 150 psig; condenser and evaporator shall be of the same construction as chiller.

The larger installations generally use water as a coolant, because of the immense amount of heat that must be drawn from the hot refrigerant in order to cool it before it goes through the expansion valve and evaporator. This cooling water becomes warm as it passes over the hot coils containing the hot compressed refrigerant. The water in turn must be cooled and this is done by pumping it to a cooling tower situated outdoors. The cooling tower is filled with wood slats over which the water trickles and through which fans draw air. The rushing air cools the trickling water by evaporation. The cooled water is sent back to the condenser to be reused. A basic cooling cycle as described above is shown here:

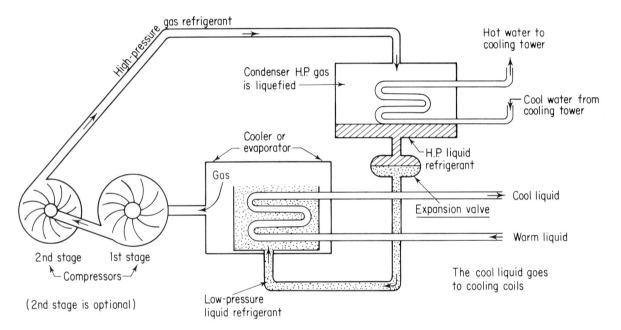

Refrigeration Cycle Using Compressor.

The specification usually calls for the cooled liquid to leave the evaporator or cooler at about 42°. This cooled liquid, which is the final result of the refrigerating cycle, performs the actual task of cooling the building. It is pumped to large coils that are surrounded by thin metal fins and look like enormous automobile radiators. The objective is the maximum transfer of heat through the largest metal surface possible. The capacity and construction of these coils are specified as follows: Coils shall be constructed of heavy gauge seamless copper tubing approximately 5/8" O. D. with copper or aluminum fins mechanically bonded to the coils. The coils shall be expanded or welded into the headers and shall be tested at 200 psi. They shall be capable of be-

HEATING, VENTILATION, AND AIR CONDITIONING

ing drained at any time. The coil bank shall rest on stainless steel structural members in a stainless steel pan which shall be connected to the drainage system of the building.

When the incoming warm, humid air flows through the cool coil it drops a good deal of its moisture content,[5] which must be drained away. It is also well to note that the parts of the system subject to constant moisture are of stainless steel, copper, or aluminum.

We have now arrived at the point where the actual transfer of heat occurs. The basic cooling cycle for large or small compressor units has been described. In the small window unit the heat transfer occurs directly in the cooling unit where the warm outside air is cooled by being blown over the cold coils and the condensed moisture is either evaporated or allowed to run free. In the large structure where the cool air must reach large and widely separated areas,[6] the cooled liquid that is the result of the refrigerating cycle must be sent to many places where it will cool the air that acts as the transfer medium. The large cooling coils mentioned above may be set at several locations convenient to the moving of the enormous quantities of air involved.

In order to obtain the utmost efficiency in this transfer of heat the cooling coils (or heating coils) are located in enclosed air handling units (cabinet type) which can be quite large and contain the coils, fans and filters and have large inlets and outlets for air. They are usually built of well braced heavy gauge galvanized steel, and are insulated against sound transmission and heat transmisson.

The fans that move the air are of two general designs: the air foil bladed centrifugal fan, or the backwardly inclined blade of airfoil form. Specifications will mention either, depending on the duty to be performed or the engineer's preference. All fans can be provided with variable inlet vanes which can be adjusted to regulate the flow of air.

Centrifugal fan blades are parallel to the inlet and outlet and literally throw the air. The airfoil blade is shaped like an airplane wing in order (as with aircraft) to lessen turbulence. The fan is built to push air, and the less turbulence, the more efficiency.

The backwardly inclined blade looks like an electric fan or propeller blade and is also shaped with a thick rounded leading edge that thins off at the back of the blade.

---

[5] The engineer has done here what is normally done by nature. A warm humid mass of air meets a cold front and as the warm air is cooled, its capacity for holding suspended moisture diminishes and the moisture condenses and falls as rain.

[6] To obtain an idea of the amount of air involved, we shall use as an example a medium sized office building of 200,000 square feet. This, multiplied by 8.6 feet which is a usual ceiling height, gives a content of 1.72 million cubic feet. (Air weighs .0749 pound per cubic foot at 70°.) This is a sealed building (windows cannot be opened by the tenants) and therefore all the air must be supplied mechanically. For reasonable comfort the air should be changed at least eight times an hour. This means that 13.76 million cubic feet or 1.03 million pounds, or 515 tons, of air have to be cooled or warmed, supplied, and exhausted every hour.

HEATING, VENTILATION, AND AIR CONDITIONING

2C1b. *Equipment sizes*

Airfoil Fan.

The sizes of all the equipment use in an air conditioning system obviously depend on the required cooling or heating load. Theoretically, to save money and space, the entire requirements of a building could be furnished by one compressor, one cooling tower, one evaporator, one condenser, and so on. This can be done very well in a small structure but to do this in a large structure would be impractical even if equipment of the enormous sizes required were made. In such a case there would be no flexibility whatsoever and one breakdown would shut down the entire system. In a sealed window building on a hot day this could be disastrous. There are buildings where this has almost been done, but it is cheap and bad design. The architect engineer who knows his business studies the building carefully. He plans the compressor sizes so that there is always a standby, which though it must be used on very hot days, is usually kept in reserve. If he has only one other compressor he may size the standby so that it can carry the building at a reasonable temperature if the main compressor breaks down. If the building has activities going on outside normal business hours, he may plan a smaller compressor to take care of this (large compressors are not efficient when they have to run throttled down).

This flexibility of equipment is a test of the architect engineer's skill and judgment. Instead of a single large compressor and auxiliary equipment as mentioned above, he could, for instance, specify five of everything for maximum flexibility, and these machines could be put on the line as the load requires. The cost of these smaller machines, however, is not in proportion to their size and such an installation would be very expensive. The architect engineer must therefore balance the varying load requirements against the initial cost of the installation and come up with an answer that will not make the owner unhappy and will provide flexibility to keep the structure in constant running order.

2C2. *Absorption Liquid Chillers*

Under Centrifugal Compressors Systems (2C1), the refrigeration cycle and the disposal of the chilled water were described, starting with the compressor and ending with the cooling coils and fans. Instead of using a compressor to start the cooling cycle, there is another method that uses heat as the source of energy and is known as the *absorption liquid chiller*. This type of chiller comes in packaged units that give a wide variety of capacities from about 100 tons to over 1000 tons. The absorption chiller has no large moving

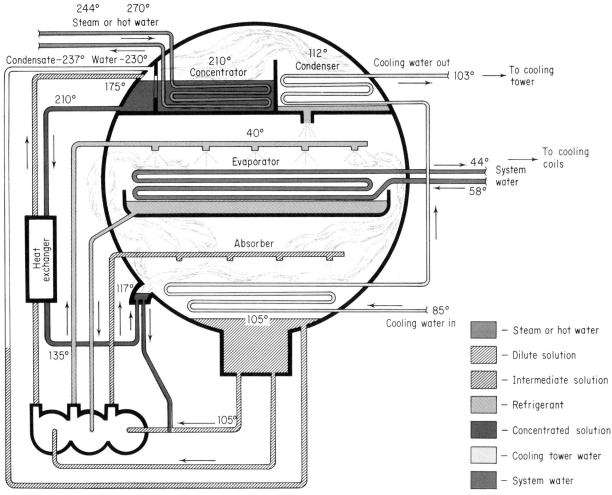

Absorption Cycle.

parts. It functions through the use of heat in the form of steam or hot water which can be exhaust steam from an electrical generating plant or waste heat from a gas turbine. A steam or hot water boiler installed for cold weather heating can be used for warm weather cooling. The waste steam or hot water resulting from an industrial process may be used to furnish the energy for such a system. Because it does not use high pressures and uses water as a refrigerant, it does not require a watch engineer. Its use should be carefully considered.

The absorption unit works on the principal that the refrigerant[7] water is made to boil at a lower temperature than the water that is being chilled. This is done by keeping the refrigerant water under very low pressures. (Water boils at a lower temperature at lower than normal atmospheric pres-

---

[7]Refrigerant water is the water that actually performs the cooling. It acts in the same manner as Freon or any other refrigerant in the compressor cycle.

sure.) The other important part of the cycle is the use of an absorbent (lithium bromide) which has great affinity for the refrigerant water. The water to be chilled passes through coils in an evaporator that is maintained at very low pressure. The refrigerant water flowing over these coils absorbs heat from the water which may enter the coils at a temperature of 54° Because of the very low pressure, the refrigerant water boils at 40° and in so doing it evaporates[8] and cools the chilled water to 44°. The evaporated water is absorbed by lithium bromide. The hot liquid that results is cooled by condenser water which flows through a cooling tower as the condenser water did for the compressor cycle previously described. The resultant cooled, relatively dilute solution of lithium bromide and water, is concentrated by the use of the steam or hot water which drives off water vapor. This is done under low pressure and therefore low temperature. The concentrated solution flows by gravity to a heat exchanger where it warms the dilute solution coming from the evaporator. The dilute solution of lithium bromide gives up its water as vapor which condenses as it passes over the cool coils containing cooled water from the cooling tower. The cool condensed water, by boiling and evaporating over the coils in the evaporator, becomes the refrigerant which chills the final working water. The cycle then starts again. The lithium bromide absorber has acted as the vehicle to first absorb the water vapor, then to carry it to the condenser where it is cooled, and then to carry it to the generator where it is concentrated. After it loses its water content in the generator, the absorber starts again by picking up the water vapor coming from the evaporator or chiller.

2C3. *Reciprocating Units*

There is a third type of equipment used for the refrigeration cycle. The reciprocating compressor, which can be used for smaller capacities such as large refrigerators, cold storage plants, and relatively small air conditioning systems, comes in fairly compact units. The compressor can be driven by a diesel engine, electric motor, or gas turbine. The multistage compressor looks somewhat like a V-type automobile engine and the banks of pistons are so arranged that the refrigerant gas is compressed through these stages to a hot liquid. The hot liquid is cooled by a condenser and the cycle continues as in the centrifugal compressor. The reciprocating compressor is most efficient for comparatively small loads and comes in preassembled packaged units having capacities of as little as ten tons and up to several hundred (although at these higher capacities other types of units should be studied). The reciprocating compressor can also run at throttled down speeds at a more efficient rate than either of the other two types of refrigerating units.

[8]It will be remembered that water absorbs heat when it evaporates. (This is known as *latent heat of evaporation.*)

2D. THE AIR DISTRIBUTION SYSTEM

The various methods of producing chilled water were described in the preceding sections of this chapter. The chilled water is pumped to cooling coils that serve to cool the air which is the vehicle used for cooling or heating. (Even in the small home, air flowing over hot radiators does most of the work, and in the window air conditioner cooled air does all of the work.) In the large structure the air must be distributed to every space within the structure and in the proper quantity consistent with the population, the activity conducted in the space, the sun and wind load, the material of the outer walls and the location of the space—whether it is on the perimeter or the interior of the structure.[9] The architect engineer must be completely familiar with all these matters. He must distribute the cooling or heating coils advantageously, keeping the air ducts as short as possible and at the same time not having runs too long for the chilled water. He must calculate sizes of cooling coils and fan units for maximum flexibility at the most economical cost per unit. (The larger the unit the more economical.) At this point one very important matter must be made clear. The primary air that is used for conditioning a building is generally kept at the same temperature winter and summer by the central units. It is cooled to about 55° when the exterior temperature is between 80° and 95° and it is warmed to 55° when the exterior temperature is between 40° and 0°. This tempered air is then further conditioned by local heating or cooling coils and by the recirculated room air. When it is between 40° and 0° outside, the air may be either heated or cooled depending on the sun and wind load. This is a rather complicated procedure, however, and as this is not primarily a text on air conditioning it will suffice here to make the reader aware of this fact.

2D1. *Zoning*

The first thing the architect engineer looks at after he has programmed the requirements and before he finally calculates the sizes of the various units is zoning.[10] In the low small structure, unless there are widely diverse activities in its various parts, this is not very complicated and can be easily taken care of. For instance, he can specify among other things larger local heating units for the north side and larger cooling units for the south side. For the tall structure however, the zoning becomes more complicated. The wind loading must be provided for. The sun loading may be different on various elevations at different heights because permanent surrounding structures may provide shade. On a fairly cold but sunny midwinter day the sunny

---

[9] This is called *programming the requirements.*
[10] Zoning is the separation of a heating and cooling system into a number of separate components, each designed to furnish sufficient air to maintain a comfort level in a certain defined area as called for by exterior weather or interior activities.

side may become so warm that it has to be cooled at the same time that the shady side is being warmed. The usual practice in the north temperate zone in a well designed office building is to separate the building into a south and east zone, a north and west zone. If it is more than 20 stories high, it should be further divided into an upper and lower zone. Sometimes it may be divided into still another zone for the first seven or eight floors. This totals at least six zones, which is a minimum requirement, or nine zones if circumstances warrant. The architect engineer can now calculate the sizes of the heating and cooling coils and fans, and locate them at the most convenient spot in each zone to take care of the weather, the business activity, and the population requirements for all year comfort.

2D2. *Discussion of Proper Location for the Plant*

The architect engineer has now determined the total and subtotal loads, and has chosen the correct sizes of heating units and refrigeration units to properly carry them, with some spare capacity. He has broken down the sizes of the units to provide the largest sizes consistent with maintaining flexibility in operation and some standby capacity. He now comes to the location of the boiler and refrigerant plant, and of the fan rooms. Shall he place them in the basement, or in the middle of the building, or on the roof? Or should he perhaps do some of each? He must weigh many factors.

2D2a. *The chiller units and heating units*

For the comparatively small building the proper location of the plant presents very few problems. A packaged cooling plant of up to 150 tons can be placed in the basement or on the street floor if necessary. A small packaged hot water or steam boiler can be located in the same place. Roof cooling towers for such small units are not necessary and the hot liquid refrigerant can be cooled by either an air cooled or evaporative condenser.[11] The air handling units can be placed immediately adjacent to the cooling units and the entire heating, cooling, and air handling plant can be placed in a small but well-ventilated space.

In the multistory building where equipment sizes are large and the spaces to be conditioned are distant, the architect engineer must give careful attention not only to the air conditioning needs but must consult with the owner, with the structural engineer, and even with a soil expert to decide where the best and most economical location for major equipment should be for the building as a whole. The considerations involved can be listed after certain assumptions are made. The assumptions are that there will be a

---

[11]The air cooled condenser cools the hot liquid by blowing air over large finned surfaces attached to the coils containing the liquid. The evaporative condenser does the same thing by means of a fan blowing air through a water spray. As the water evaporates it cools the hot liquid refrigerant. It is really a miniature cooling tower.

HEATING, VENTILATION, AND AIR CONDITIONING

cooling tower on the roof; that in order to obtain as pollution-free air as possible the major air handling equipment will be on the roof, and that soil and foundation conditions are such that there is a free choice between basement and roof. The advantages of placing a plant in the basement are:

1. The heavy equipment loading of compressor, condenser, and boiler plant will be low in the structure and will require no strengthening of the structural frame.
2. The problem of vibration and sound will be minimized.
3. Heavy equipment may be replaced more easily.
4. The utility service is immediately accessible to heavy equipment.
5. If there is some delay in the completion of the structure due to strikes, fire, or other eventualities, it may be possible to place the system into service earlier if the cooling tower and condenser water system can be utilized.

The disadvantages are:

1. The possibility of excavation or basement waterproofing problems.
2. The boiler stack and the condenser water lines must be run to the roof, thereby occupying valuable rentable area on each floor as well as being costly. This also holds true for the chilled water that must be pumped up to roof fan units.
3. The plant will occupy rentable area in the basement.
4. The ventilation of the below-grade space is necessary.
5. Maintenance will be divided between roof and basement equipment.
6. The controls become more involved.

To place a plant on the roof the advantages are:

1. The condenser water and the chilled water lines are at a minimum because the cooling tower and most of the fan units are immediately adjacent. This saves installation expense and minimizes heat losses.
2. The boiler stack can be eliminated.
3. No rentable space is lost because the roof space occupied by the plant is not rentable and the shaft space is eliminated.
4. Centralized maintenance and controls.

The disadvantages are:

1. The heavy roof load requires extra structural frame design.
2. The utility services, either gas or electricity, must be extended to the roof.
3. The vibration and noise problem become more complicated.
4. Heavy replacement equipment will have to be hoisted to the roof.
5. The roof plant is the last installation and must wait for the completion of the structural frame, structural floors, and in bad weather the completion of the building skin.

Today more and more high-rise buildings are placing their plants on the roof.

HEATING, VENTILATION, AND AIR CONDITIONING

2D2b. *The fan rooms*

The location of the fan rooms is a product of the zoning and of the building activity. If the tenancy requirements on lower floors are heavy, then fans may be placed in the basement or midway in the building. If the requirements are distributed evenly through the building, then fan units may all be placed on the roof or may be divided at a mid-floor to feed down from the roof and up and down from the mid-floor. Where business on street or mezzanine floors is conducted outside normal business hours the architect engineer may provide chilled water to fan units located within these areas or may provide small packaged units to accommodate their entire requirements. There must be as many fan units as there are zones.

2D3. *The High- and Low-Pressure System*

Once the fan rooms are located, the system of distribution of the air must be determined. There are, of course, several methods of doing this. A fairly conventional way of distributing air in a northern climate with severe winters is by means of a high-pressure perimeter air and low-pressure interior air system. In this system small diameter air ducts are run up the perimeter of the building alongside the exterior columns and carry warmed or cooled air at a fairly high velocity. This air is carried to induction[12] perimeter units that further heat or cool the air and then release it into the space. The perimeter units can provide conditioned air for a space about 20 feet in from the perimeter. In the induction unit, some of the air is recirculated and some comes in fresh from the ducts. The excess air is exhausted by means of ceiling louvres which, by means of ducts or a plenum (see Section 2D7) ceiling, return it to main exhaust ducts. After being recooled or reheated and mixed with fresh air, it is sent back to perform its function again.

For the interior conditioning, large ducts carry air at low velocity to ceiling louvres or diffusers, which to quote a specification "shall provide required air throw and spread, with no apparent drafts or excessive air movement within the ventilated or air conditioned area." These diffusers, which are used in every system of distribution, may be round or square or rectangular and can be fixed to blow air in all or in specified directions. They can also be linear (a straight and narrow slit with vanes arranged so that the conditioned air does not blow straight down). Another way of distributing the interior air is by means of lighting fixtures that are manufactured with a metal jacket superimposed on the fixture. This jacket receives air from a duct and transmits it downward into the space by means of slits on the sides of the fixture.

[12]Induction units to be explained later.

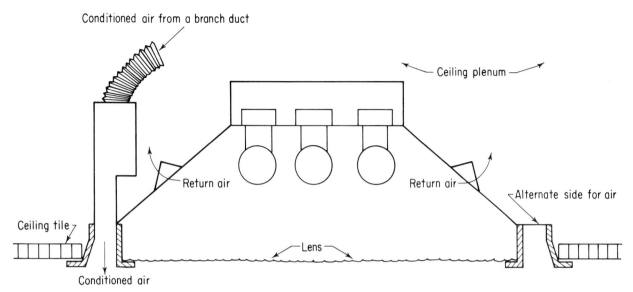

Air Distribution Through a Lighting Fixture.

Interior systems do not require much heating except in very severe weather, since they are insulated from the exterior by the curtain of warm air provided by the exterior conditioning. In the summer, because the interior gets no sun load the entering cooled air is more apt to have to be warmed. To do this the interior zones are provided with tempering coils that contain either hot water or steam and over which the cooled air is blown. The flow of air over the coils is regulated by thermostats.

2D4. *Single-Duct System*

In small buildings, where zoning and flexibility are not important and the activity within the building is generally the same everywhere, the single-duct system with terminal reheat may be used. The air is sent up to the floors by means of two vertical ducts, one for the south and east elevations and one for the north and west elevations (in the latitude of the northern part of the country). The air is sent through a tempering coil that is thermostatically controlled and then distributed by means of a duct that sends air into every space by ceiling diffusers. In winter, extra heat is furnished by under-window radiation. In the summer, depending on outside temperature and the requirements of the thermostat, the coils at each floor main outlet either warm the air or allow it to come into the spaces at 55° where it is mixed with room air to temper it.

HEATING, VENTILATION, AND AIR CONDITIONING

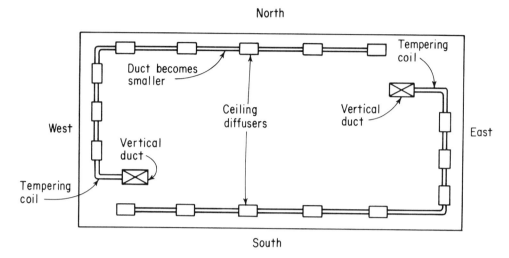

Single-Duct System, Small Building.

2D5. *Double-Duct System*

In milder climates, a very popular means of air distribution is the double-duct system. This system consists of a complex of main vertical and branch horizontal ducts that carry high-pressure warmed air and a parallel series of ducts carrying high-pressure cooled air. The horizontal ducts terminate in mixing boxes located in the ceilings in every area that is controlled by a thermostat. The mixing box, which is just what its name implies, mixes warm and cold air in proportions called for by the thermostat, and after reducing the air velocity it sends the air into the specifically controlled space by means of smaller auxiliary ducts. A mixing box specification states that the box shall be constructed of a minimum of 22 gauge galvanized sheet metal, shall be of bolted construction, and shall have all joints sealed. The dampers controlling the flow of air into the box shall be leak proof and the volume of the box shall be such as to provide for the necessary pressure drop. It must also be sound insulated.

This system is fed from large vertical ducts in the interior core and provides for both the exterior and interior zones. The exterior is supplied by means of diffusers located in the ceiling over the window openings or by under-the-window plenums. These diffusers are set to blow a curtain of warm or cooled air over the window depending on the requirements of the weather.

Another means of distributing air through double ducts is by means of an air floor. This is a structural cellular floor that is adapted to carry air as well as electricity through its raceways. The air cells are, of course, much larger than the electrical cells. The air cells are connected to the vertical risers by means of supply header ducts at each floor and carry the warm and

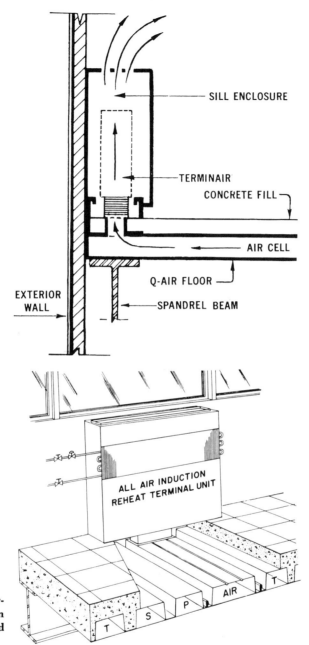

Robertson Q Air Floor. The Terminal Unit Can Be an Induction or a Fan Coil Unit or Can Be Fed by a Double-Duct System.

cooled air to mixing boxes from where it is distributed at the perimeter by outlet boxes usually situated under the windows, or distributed to the interior by ceiling diffusers.

It was stated at the beginning of this section that this double-duct system is most suitable for mild climates. This is so because in cold weather the amount of warm air that would have to be blown down over the win-

HEATING, VENTILATION, AND AIR CONDITIONING

dow surfaces would create uncomfortable drafts, and unless blown at fairly high velocity this warm air would not completely curtain the glass because the cold glass would cool it too quickly. The ducts themselves would also have to be extra large. If the double-duct system is used in cold climates, there should be auxiliary heating in the form of base radiation at each window.

2D6. *Perimeter Window Units*

In distributing the conditioned air the architect engineer must pay particular attention to the perimeter areas. With exterior walls and glass that transmit heat in either direction, he must see to it that enough air at the

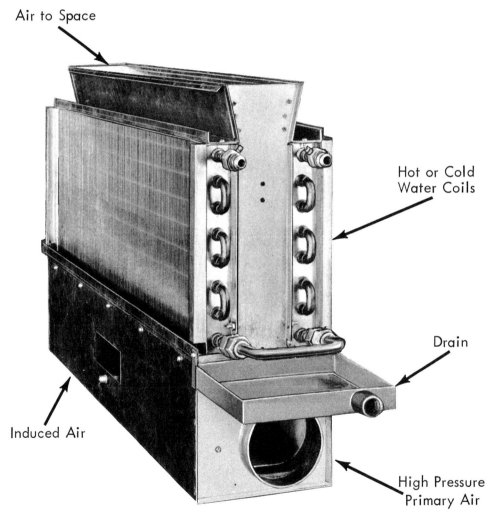

Perimeter Air Induction Unit.

correct temperature is supplied so that the occupants of these areas will be comfortable and the air will form a protective curtain for the interior area. In the conventional exterior high-pressure system the planner will specify induction window units. These units consist of a series of coils that can carry either hot or cold water. Thin metal fins are bonded to these coils so that the maximum metal surface is presented to the air that blows over them. The air that comes from the vertical high-pressure riser ducts enters a large chamber under the coils (the plenum) where its pressure is reduced. As this fresh warm or cool air blows over the coils, its velocity induces room air to enter into this plenum and mix with it. The room air is thus freshened and reheated or recooled. The flow of air is controlled by a damper and the flow of liquid in the coils is controlled by a thermostat. A specification for such a unit would state: The coil shall consist of ½-inch O. D. copper tube hydraulically bonded to aluminum fins; coils shall be provided with manual vent cocks; air plenum shall be of galvanized steel and shall be fitted with a damper for adjustment of the primary air; the unit shall be furnished with condensate pan of corrosion resistant finish; the unit shall be so constructed that if the quantity of the air delivered and the water temperature are in accordance with the design no condensate will form; all units to be furnished with lint screens. The enclosure for the unit with its grille shall be furnished by other subcontractors.

Another popular window unit is the fan coil unit. This often comes as a completely packaged unit that looks like a radiator enclosure. It contains a coil of copper tubing bonded to aluminum fins and like the induction unit can be used for circulating hot or cold water. Unlike the induction unit, however, it blows air over the coils by means of fans situated under them and is not directly connected to a source of fresh air. The fresh air in this case is supplied by the interior system which delivers it through ceiling diffusers or lighting fixtures as the case may be. The installation of a fan coil system is generally considered less expensive than the installation of the air induction type because the exterior high-pressure risers are eliminated. However, the possible maintenance of an individual motor for each unit must be considered. In addition, because there is no outside air delivered directly to the unit, the interior air system must be made larger and the ducts supplying this air must come closer to the perimeter spaces. The fan coil unit is more economical only in apartment houses or small buildings, where there is no objection to an exterior grille at every unit to furnish fresh air. This really makes it a conventional window unit such as we all know, except that the hot or cold water for its coils is centrally heated or cooled.

2D7. *Exhaust System*

All systems generally exhaust their air in the same way, which is by a ceiling plenum reinforced by exhaust ducts running in the hung ceiling space. A ceiling plenum is the space between the underside of the floor

HEATING, VENTILATION, AND AIR CONDITIONING

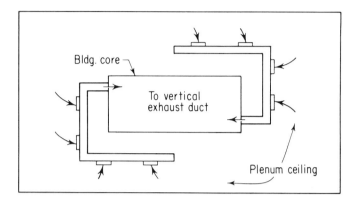

Collection Ducts for Exhaust Air in Plenum Ceiling.

above and the top of the hung ceiling (see Chapter 14). The exhaust air comes into this space by means of ceiling louvres. It is pushed in by the incoming conditioned air and is helped along by exhaust fans that exhaust it from the ceiling into ducts running part way into the ceilings and connected to vertical risers. The exhaust from toilet rooms also sucks air out of interior spaces. Special use rooms, such as conference rooms and other areas where many people are gathered, have specially large exhaust grilles running into the plenum or, if the load is heavy enough, the grille may connect directly into an exhaust duct.

2D8. *Sound Abatement*

All architect engineers are aware of the problem of sound in designing an air conditioning system. The compressors themselves must not only be dynamically balanced to minimize vibration, but they must also be set on bases designed to dampen and isolate any vibration from the structural frame. Usually, the large machines are mounted on heavy concrete "inertia" bases which through their mass, dampen vibration. In addition, these bases are set on waffle neoprene pads of specified hardness and thickness. All fans are isolated by means of heavy springs mounted on hard rubber pads. Pumps, motors and piping carrying chilled water or condensate are isolated by means of spring mounted ceiling hangers on horizontal runs. On vertical pipe runs where they are supported on every floor by the structural frame, such supports are isolated from the frame by means of pads of neoprene or hard rubber.

The most difficult sound to control is the sound of the air itself as it is sent to the various areas. The enormous quantities of air that are necessary in a building of any size were previously mentioned in this chapter. The air must be transmitted at the highest velocity possible so that the smallest possible duct size may be used. Such maximum air velocities in ducts are set by code or by ASHRAE[13] standards. In a high-pressure system, a velocity of 3000 fpm may be allowed for a main duct and in a low-pressure system this would go down to 1500 fpm. Branch ducts are permitted less velocity. This means that much care must be exercised to prevent the air noise of ducts, fan rooms, window units, and mixing boxes from annoying the occupants of a building. The specification for the fan room where the air is cooled or heated by pipe coils and then blown through the ducts, first concerns itself with the fan itself which must be statically and dynamically balanced and mounted on rigid steel shafts carried in self-aligning bearings. The fan is mounted on steel and hard rubber springs. The air noise is taken care of by specifying that the surrounding steel casing be lined with fiber glass. The ducts themselves are fitted with duct silencers, which are metal casings filled with mineral or glass fiber and inserted into the duct for at least a 20-foot distance from the fans. A specification will provide for testing of such silencers at an air flow of 2000 fpm and will state the maximum allowable sound transmission in decibels. All ceiling diffusers are required to keep within a certain decibel rating at specified air flows. The mixing boxes that receive high-pressure warm and cold air and lower its velocity must be carefully sound-proofed. The specification calls for the interior of the metal casing to be lined with 1-inch-thick, dense coated glass fiber. Again sound tests are called for. The plenum chambers of induction units are sized and constructed to prevent noise. In spite of all these precautions strange things happen and weird noises occur, and the architect engineer must be prepared to cope with them. Recently, a well designed building was put into operation and the fans were started. Within a very short time several squad cars pulled up to the building and the police wanted to know what was causing a strange howling noise that was disturbing the neighborhood for blocks around. It was traced to a large fan assembly on a middle floor. Apparently (and no one is sure), the speed of the fan set up a sympathetic vibration in the metal casing. The entire casing (10′ × 10′ × 15′) had to be double blanketed with acoustic material. Cost?—$8000!

2E. The Control System

The air conditioning control system must be able to maintain predetermined temperatures and humidities during all seasons and for every work activity. This is a highly complicated task and requires a control system that works almost like a computer. The control system performs this task by first measuring the changes in temperature and sometimes humidity, in the various areas; second, by converting the observed changes into energy,

---

[13] ASHRAE stands for American Society of Heating, Refrigerating and Air Conditioning Engineers.

HEATING, VENTILATION, AND AIR CONDITIONING

either pneumatic or electrical; and third, by using this energy to correct the system by means of controlling devices in order to bring it back to the predetermined levels. Control systems are activated either pneumatically or electrically. A pneumatic system consists basically of a source of compressed air (by means of an air compressor) that provides the operating energy. This compressed air is run through copper tubing to all controlling devices such as thermostats, dampers regulating air flow, humidity and pressure controls, and other devices. The controlling may be accomplished by means of a thermostat calling for more heat and thereby opening a valve to allow more hot water to flow through a coil. The actuating device in this cast may be a bimetal strip or a sealed bellows that will open or close a valve controlling the compressed air flow which opens or closes the water or steam valve. Humidity is controlled by a moisture-sensitive material that again actuates a valve. The pneumatic control system uses relatively simple devices that do not require a great deal of maintenance. It may be more economical than the electrical control in any area where code requirements call for low-tension wiring to be run through conduit. It can be used in places where explosion hazards exist.

The electric control system does the same work as the pneumatic, except, of course, that the flow of electric current rather than the flow of air actuates the controlling valves. The electric control system has the advantage of being able to amplify relatively feeble impulses from the sensing devices and the electric wiring is more reliable for remote control of the system.

A general specification for a temperature control installation would state: The temperature control system shall be of the pneumatic (or electric type) and shall include thermostats, automatic valves, damper motors, duct thermostats, electro-pneumatic valves and switches, and all air supply equipment and piping required to maintain the conditions described. The specification goes on to describe the sensing devices and controls and all the other devices that are necessary to keep the interior areas at a level temperature. The complexity of the system can be understood when it is noted that its duties include keeping air and the water in tempering devices flowing in sufficient quantity to keep the sunny and shady sides of a building at the same temperature mean, while maintaining stated temperatures in widely different activities. One can imagine the constant movement of hundreds of valves, and dampers, and pressure regulators as they respond to the sensing devices that are constantly trying to attain the temperature specified by the architect engineer.

2F. WATER TREATMENT AND FILTERS

All water used in air conditioning systems must be treated to prevent corrosion and the deposit of any scale. The chilled or hot water travels through hundreds of miles of small-diameter tubing, and any blockage of the lines or their corrosion would be disastrous. The condenser water that then flows over hot tubes in the condenser, must not leave any deposit on these tubes, otherwise the entire refrigeration cycle will be jeopardized. To prevent this, the architect engineer specifies a chemical feed system that tests the water and automatically injects the proper amount of corrosion inhibiter into the water. He also controls the pH[14] content of the water. In addition to the automatic controls, the specification will call for a supply of test kits so that the operating personnel may double check the efficacy of the controls by taking periodic tests. The architect engineer must know the chemical makeup[15] of the local water and the rate of feed, and the type of chemical must be specified exactly to neutralize these conditions.

Filtering the supply air in an air conditioning system is as important as treating the water. Today, when smog and industrial haze blankets many cities, the use of filters that will remove impurities from the air is of the utmost importance. The comparatively pure air that can be produced by good filtering will not only keep employees happy but can help a great deal in reducing building maintenance costs. In research laboratories, operating rooms, and plants where delicate electronic parts are assembled, the filtering must be absolute. The filters normally provided for a structure where office work is the major activity are of the "high-efficiency mechanical type" and are specified as follows: furnish and install in locations shown prefilters and high-efficiency filters; prefilter section shall be of 2-inch throwaway type and shall have an average efficiency[16] of 68%; the high-efficiency filter cells shall consist of galvanized steel peripheral frames supporting multipleated submicron[17] diameter glass fiber media. The average arrestance efficiency when tested in accordance with National Bureau of Standards "dust spot" test, using atmospheric air, shall be 85%. Many specifications demand an efficiency of from 90 to 95%.

There is another popular type of filter known as an electric precipitator. In this filter, high-voltage electric wiring is used to charge dust particles in the air that flows over the filter. The dust is charged by passing over wires carrying from 5000 to 20,000 volts, depending on the size of the installation. These particles are then precipitated on a surface of opposite polarity. This surface can be washed mechanically at certain intervals that

---

[14]The pH factor is the ratio of acidity to alkalinity in water. A factor of seven represents the neutral point. Lower than seven represents an acidity which increases the corrosiveness of the water. A corrosion inhibitor adds alkalinity to the water and either keeps its factors at seven or slightly above.

[15]This includes the alkalinity or acidity, the dissolved CO_2, the dissolved oxygen, any dissolved mineral salts, sulphur, etc.

[16]The efficiency of a filter is measured by the amount of atmospheric dust it is able to filter out. In this case, 68%.

[17]A micron is 1/1000 of a millimeter. A millimeter is roughly 4/100 of an inch. There are about 25,000 microns in 1 inch.

HEATING, VENTILATION, AND AIR CONDITIONING

are measured by an air pressure drop which tells when the filter has become dirty. Such filters are specified at 90% efficiency by the N. B. S. discoloration or dust spot test. A throwaway prefilter is also usually called for.

It was mentioned above that some areas of activity require absolute filtering. Such filtering is done by mechanical filters which are made of extremely fine fibered paper that is closely pleated and held in a rigid frame. Some filters of this type have been tested to hold back 99.95% of particles measuring .3 micron in diameter. This filter can therefore pick up all solid particles down to 1/80,000 inch in diameter—including many bacteria mold spores and all dust.

There are other essential portions of an air conditioning system such as valves, strainers, piping, the gauges of the metal required for certain size ducts and so on. This chapter has, however, confined itself to the general theory and practice in this field. It has pointed out the requirements for a system that will provide flexibility, and interior weather that will keep occupants satisfied, while at the same time satisfying the owner by an economical installation and operating cost.

3. THE "ALL ELECTRIC BUILDING"— A VARIANT FROM THE CONVENTIONAL

In certain parts of the country where the cost of electricity permits it, the architect engineer has been designing an "all electric" building. This means that all heating and cooling is done by the use of electricity. There are various examples of this method, which vary in their utilization of electrical energy.

In one structure now being erected, the system is built around the utilization of the heat of light. The air in the building is distributed through a conventional dual duct system that runs through the hung ceiling and which, through mixing boxes, sends air up to the floor above at the building perimeter by means of very low sill enclosures. The basic difference between this building and the conventional building is that all of the 9000 six-tube fluorescent lighting fixtures are water cooled. Water is circulated through the shell of the fluorescent fixtures to extract a substantial amount of the heat of the light and by means of piping removes this heated water to where it can most efficiently be used. A major benefit of the water cooled fixture is in removing this heat of light to reduce the summer cooling load in the air conditioned space. This results in diminished air quantities, decreased size of air handling equipment and ducts, and reduced floor to floor height. The engineers have estimated that one full story height of a 24-story building

An All Electric Building's Heating and Cooling Cycle. ⟶

HEATING, VENTILATION, AND AIR CONDITIONING

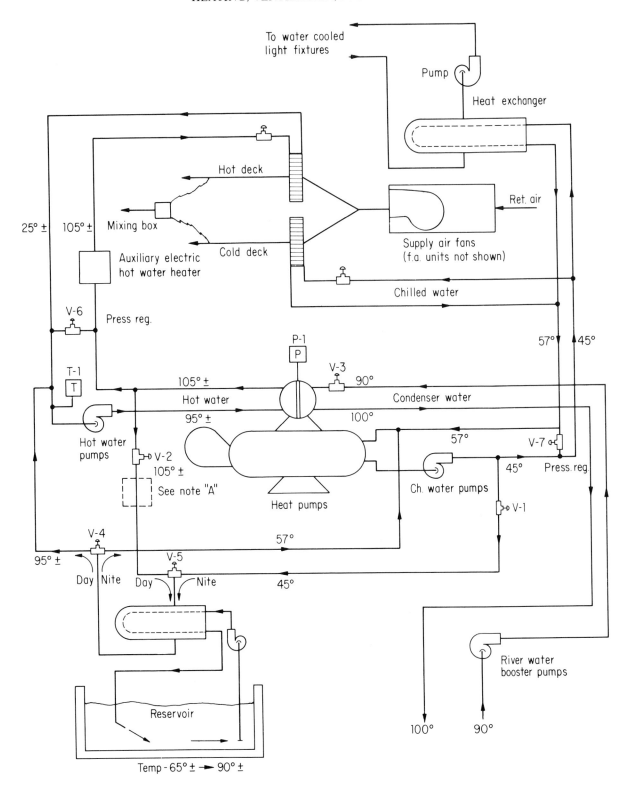

Note "A" - Location of possible domestic hot water preheater

HEATING, VENTILATION, AND AIR CONDITIONING

has been saved. Another major benefit is that the heat extracted by the water cooled fixture can be used to temper the conditioned air, where this is required, or to heat the building during the winter. When the warmed water is not needed it is sent down to a 150,000-gallon reservoir, situated in the sub-basement of the building, where it is stored until it is required. In winter this warm water can be used for keeping the building heated to normal off-hour temperatures during weekends and nights when lights are off and the building is uninhabited.

Although the six-tube $2' \times 4'$ lighting fixtures are built to maintain 150 foot-candles of light, there is not sufficient heat of light generated to provide the building's heating needs during extreme winter conditions. After the heat of light is extracted from the water by means of two refrigeration machines that are used as heat pumps, electrically heated hot water boilers can supplement the heating requirements at peak periods. These hot water boilers also supplement the available heated water in the reservoir when required. An additional feature of this building is that all windows are double glazed to provide the utmost efficiency.

In another example, electricity is used to drive centrifugal compressors. The plant is located on the roof, and instead of a cooling tower for cooling the condenser water, the architect engineer has specified large air cooled condensers. These are cooled by large electrically driven fans. The system becomes all electric when the chilled water resulting from the refrigeration cycle is circulated through the building and is sent to a fan room on each floor that transmits 56° air at high velocity to plenum boxes where the air is heated electrically to the temperature called for by the zone thermostats. The warmed exhaust air is returned through a plenum ceiling, where by heat exchange it tempers the high-velocity air that is on its way (in ducts) to the plenum boxes. In the winter additional heat is furnished by perimeter electrically heated units under the windows.

A knowledgeable architect engineer team can add many refinements to the weather conditioning of a building at no great initial or operating cost. For instance, in very cold weather, when the outside humidity is low (because water vapor has been condensed below the dew point), interiors can sometimes be uncomfortably dry. The small home can use a humidifier. In the large building, moisture can be injected into the supply air by controlled steam jets or by sending the air through a water spray. The cost and maintenance of this refinement should be investigated and if possible, it should be included. Another refinement is to make the main supply ducts large enough to be able to supply the building with 100% fresh air on a temperate spring or fall day. This fresh air supply can also be used on a 40° sunny day when the sunny side of the building becomes warm and should be cooled. Such a day may occur in mid-winter when the cooling tower is not working and the refrigeration compressors are down. In such a case the warm sunny side can be flooded with 40° cool air to make it comfortable while the shady side is being warmed by heating coils.

Weather conditioning is interesting and stimulating and calls for thoughtful and knowledgeable ingenuity.

CONTENTS

1. THE PLUMBING INSTALLATION
   A. Basic Code Requirements
   B. Determination of Building Requirements
      1. The Sanitary Facilities and Water Supply
         a. Drinking Water
         b. Chilled Water
         c. Plumbing System Design
      2. A Typical General Specification
      3. Materials and Methods of Installation
         a. The sanitary waste system
         b. The storm drainage system
         c. Sewage ejectors and sump pumps
         d. Cold water systems
         e. Water treatment
         f. Hot water system
         g. Fixtures and accessories
         h. Plumbing tests
   C. The Distribution System
   D. Process Water and Industrial Wastes

2. FIRE FIGHTING SYSTEMS
   A. Standpipes and Pumps
   B. Sprinkler Systems

10

PLUMBING AND SPRINKLERS

The installation of an efficient and safe plumbing system is a basic requirement in any structure occupied by people as a residence or as a place of work. This system supplies water for drinking, cleaning, washing, sanitary facilities, air conditioning, fire protection[1] and laboratory and industrial processes. It also provides for the effectual removal and disposal of the wastes from these uses.

1. THE PLUMBING INSTALLATION

1A. Basic Code Requirements

Building codes recognize the importance of the plumbing system and are specific as to its installation. In a private residence, the code specifies sizes and materials of piping for water supply lines, for waste and for vent lines. It will detail how a waste line is to be sloped and how a vent line is to be connected. The diagram below shows a typical basic sanitary plumbing connection. The trap, which is filled with water, is located where it is in order to seal off the waste lines from the interior spaces. The vent line which opens to an outdoor space allows the escape of any trapped gases. The minimum facilities required by a typical code are as follows: In each private dwelling or family unit in a multiple dwelling there must be at least one water closet,

---

[1] Fire protection is afforded by standpipe and sprinkler systems.

PLUMBING AND SPRINKLERS

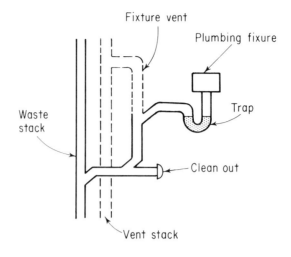

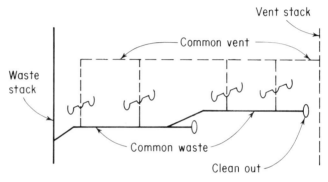

Basic Plumbing Fixture Connection.

one lavatory (or wash basin), one kitchen sink, and one shower or tub. In hotels or dormitories there must be at least one water closet, one lavatory, and one shower or tub for each six occupants of each sex. Requirements for industrial, office, and other buildings are also specifically spelled out.

Because the proper disposal of wastes is so important to health, the building codes in any sizeable community detail how a connection to a public sewer must be made. In the absence of such a system, the code goes into detail on septic tanks and drain or leaching fields. In specifying the sewer connection, the code states the kind of pipe to be used, depending on the soil through which the pipe runs, and the diameter of the pipe, depending on the number of fixture units connected to the system. Where no sewer is available the code is strict in its requirements. It usually requires public health official and building department approval. The code calls for percolation tests which determine the absorptive capacity of the soil and from this the authorities will specify the size of the drainage field. In outlying suburbs without public sewers the architect engineer would be wise to have a test pit dug when planning even a single-family house (as mentioned in Chapter 5). The underlying soil encountered will give him an idea of the drain field required. This can be expensive and should be anticipated. It can be especially expensive when a large structure with heavy potential population is built in an outlying area. When this occurs the architect engineer should employ an expert in waste disposal. In many cases the underlying soil may be incapable of absorbing the amount of effluent discharged by the septic tanks and the planner and owner may be faced with providing a sewage disposal plant whose effluent can be safely discharged into a natural body of water. With the entire community now becoming increasingly aware of water pollution, the importance of this cannot be overemphasized.

1B. Determination of Building Requirements

1B1. *The Sanitary Facilities and Water Supply*

In large structures such as places of public assembly, industrial plants or office buildings, codes specify sanitary facilities as determined by the number of occupants. The planner of an office building, depending on the code, will calculate the occupancy by allowing 1 person per 100 square feet of area. He will figure that normal male and female occupancy is equal. He notes the code requirements for each sex and plans the toilet facilities on each floor. The farsighted architect engineer will make the plan flexible. He will place male and female facilities side by side with a semimovable wall between them so that if the population of any floor becomes predominantly male or female, the dividing wall can be moved.

In planning toilet rooms in office buildings, the architect engineer must consider comfort and convenience as well as providing the bare necessities. He may wish to plan the number of wash basins to exceed code requirements. The location of mirrors and makeup facilities in ladies rooms, while not strictly plumbing, certainly bears a relationship to it and should be pleasingly arranged. The writer has frequently encountered ladies' committees who took exception to installations providing too few basins or too narrow compartments. If the structure is for a single occupant or for several large occupants it is recommended that the architect engineer consult beforehand with a ladies' committee.

This applies also to the industrial plant. With changing work patterns and Fair Employment laws, preparation should be made for the future expansion of female help. Here the washing facilities are also extremely important. The planner should bear in mind that the blue collar worker is constantly engaged in raising his status and the facilities supplied to him should reflect this. A thorough study should be made of the most modern facilities available for washing and showering in connection with locker rooms. The initial expense will not compare to the possible future expense necessary to meet workers' demands.

PLUMBING AND SPRINKLERS

Multiwash Basin, Multishower Stall. Bradley.

In addition to providing facilities for the occupant, the planner must also furnish facilities for other uses in the building. Space must be provided for slop sinks and other cleaning facilities. Hose connections have to be provided for washing exterior areas. Water must be provided for the air conditioning system and the heating systems, and for fire fighting.

1B1a. *Drinking water*

In any discussion of the plumbing system it is important that the planner have a thorough understanding of the requirements for potable[2] water. Where a structure is planned away from a public source of water it must be provided with a private source of clear uncontaminated water by means of a well, or a stream, or another body of fresh uncontaminated water.

Depending on the underlying strata, wells may have to be drilled down as far as several hundred feet. The holes are cased with metal pipe and the system is provided with a pump and a storage tank which holds the water under pressure. The water from such sources should be tested by public health authorities before use and at reasonable intervals. The well should be situated away from any possible source of contamination. Codes call for a distance of at least 75 feet. Even with the utmost care there are cases where a fault in rock strata may cause contamination as deep as 100 or more feet below grade, which is why such testing is necessary. The author knows of an instance where several wells more than 100 feet deep

[2]*Potable* is a term used by codes and specifications to describe water that is safe to drink.

PLUMBING AND SPRINKLERS

were contaminated by detergents whose source was unknown. Building codes are specific about the supply of potable water. In many areas where process water, or condenser cooling water for air conditioning systems, or other cooling water is obtained from possibly contaminated sources, all codes state that the potable and nonpotable water be clearly identified. This is done by marking them at the point of supply and by color coding the piping so that there is no possibility of cross connecting such water into a drinking water supply. Contaminated water is always very scrupulously controlled.

1B1b. *Chilled water*

A supply of chilled water for drinking purposes is obviously a necessity wherever people gather. In places of public assembly a code may call for one drinking fountain for every 1000 people or at least one per floor; in schools, at least one per floor; in factories at least one per floor per 100 workers. Code demands should generally be exceeded by practical considerations and regard for the convenience of the occupants. Drinking water can be chilled at a central point and distributed, or individual drinking fountains containing refrigeration units may be used. The initial installation and maintenance of a central system is more expensive[3] than individual refrigerated drinking fountains and most buildings chill the drinking water at the point of use.

1B1c. *Plumbing system design*

In the design of a plumbing system as in every other facility for a building the architect engineer must keep the flexibility of the system in mind. He must allow for increased future uses for water for food preparation, for cooling, for sanitary facilities, for fire protection. He starts by slightly oversizing the supply and waste lines throughout the structure and also at the source of supply or discharge. In a private residence such foresight will be appreciated in the event of an addition. It can help the sale of a house. In the industrial plant an oversized sewage disposal system or extra large supply lines, even from a distant source of water, can save great expense when expansion is planned. In an office building the architect engineer should provide for future use by providing empty vertical water supply lines and waste and vent lines at one or more locations on a floor remote from the central core where the building facilities are always located. Such "wet columns," as they are called, are extremely important when a tenant requires a private toilet, or extra drinking fountain, or special food preparation facilities, and can also be an important lease consideration.

---

[3]If drinking fountains are located one above another it is sometimes cheaper to use a central system.

PLUMBING AND SPRINKLERS

1B2. *A Typical General Specification*

The following typical plumbing specification for a large structure will consolidate the preceding sections. The specification calls for: Sanitary waste systems and drains; storm drainage systems; sewage ejector and sump pump units; domestic cold water system with pumps and controls; domestic hot water system with hot water generators and circulating pumps; plumbing fixtures and fire fighting piping and equipment. The plumbing specification also calls attention to work that must be done in connection with the air conditioning, water supply, and waste system.

1B3. *Materials and Methods of Installation*

1B3a. *The sanitary waste system*

Generally good practice for the waste system is to use extra heavy cast iron soil pipe above and below ground within a building, and extra strength vitrified clay pipe outside the building. Horizontal sanitary piping is usually laid at a slope of ⅛ inch per foot of length and sharp bends should be avoided. Vent piping is usually of galvanized iron up to 2½ inches in diameter and of standard weight cast iron above this size. It can also be of steel or wrought iron.

Specifications and codes are specific about vent sizes and how they are to be connected to the various fixtures. When detergents became common, the suds from them were likely to fill waste lines in apartment buildings and back up into the lower floor fixtures. Specifications and codes now call for stack offsets and relief stacks to avoid this condition. Stack offsets are also called for in buildings over a certain height. This is an example of the forward looking planner working with the code to prevent a possible health hazard.

1B3b. *The storm drainage system*

Storm drainage systems are separate from the sanitary waste system for a very definite reason. In heavy rains storm water can cause a sanitary system to overflow, with serious consequences. Codes prohibit storm drains from connecting with a sanitary sewer system except under special conditions. The architect engineer specifies storm drainage for all open areas and plans the size of the discharge piping from the size of the drainage area. The planner should familiarize himself with the local climate. If very heavy rainfall is normal, the drainage system must be sized accordingly. Storm drainage piping may be galvanized or cast iron. The slopes of the various drainage areas should be carefully checked and the area drains specified to meet local conditions.

PLUMBING AND SPRINKLERS

1B3c. *Sewage ejectors and sump pumps*

In cities where land is valuable, there are below-grade basements in most large structures containing sanitary facilities that are below the level of the city sewer system. These basements may also have problems with the seepage of ground water. To alleviate this situation the architect engineer specifies sewage ejector and sump pump units which are actuated by float valves and have the capacity to dispose of all waste and ground water into the city sewers. Such units are specified to have either special iron or bronze impellers, waterproof motors, and other high-quality fittings. They are subject to heavy use in wet confined spaces.

1B3d. *Cold water systems*

Water systems usually use bituminous coated cast iron piping for the service piping between building and street mains where the service lines are over 4 inches in diameter. All pipe and fittings are specified to conform to Federal or American Water Works Associations standards. The distribution piping inside the building is specified to be either hard drawn copper tubing (type K or L) or 85% red brass. The fittings are specified to be brass or copper and the solder often is silver. Water piping must be run without sudden dips or bends so as to avoid traps and should be provided with drains at all low points. Most codes and specifications require that provisions be made to avoid water hammer, which is caused when a valve is closed suddenly. This can be done by providing air chambers at the end of a line or by mechanical shock absorber devices. Long straight lengths of water piping should have a shock absorbing device at the end of the run. At high pressures a serious water hammer can rupture a fitting.

Before any water system can be used it must be tested, cleaned of all oil, grease, and other foreign matter, and sterilized by slowly filling the system with a mixture of chlorine and water (50 parts per million). After waiting at least 24 hours (200 ppm for 3 hours) the system can be flushed and is ready for use.

1B3e. *Water treatment*

The corrosive effect of water can present a very serious problem in the maintenance of a building. All water contains dissolved oxygen and carbon dioxide as well as various dissolved minerals. Because "soft water" contains only very minute quantitites of dissolved minerals which might deposit a protective coating on the insides of pipes, it is apt to be more corrosive than hard water. The dissolved carbon dioxide in the water combines with water to form carbonic acid, which is the cause of most of the corrosion in iron or steel piping. The dissolved oxygen also attacks iron or steel through an electrochemical process. This is why copper or 85% red brass piping has come into widespread use for the distribution systems. Copper is much more resistant to the action of the free oxygen or carbon dioxide, than are iron or steel even when galvanized. There are water treatments available for lowering the acidity of water. These consist of alkaline chemicals such as sodium silicate or hydrated lime which increase the pH[4] value of the water and reduce its corrosiveness. There are also treatments for the hardness in water. Such "hardness" is caused by dissolved salts such as calcium and magnesium carbonates or sulphates. These dissolved salts are concentrated in steam boilers as the water boils off and the salts are deposited as scale on boiler surfaces. Such scale can cause tube rupture because scale deters transfer of heat to the boiler water and may cause overheating and consequent weakening of the tube wall. There are several methods available for treating the dissolved salts that cause hard scale, such as the Zeolite softener or lime–soda ash, either of which acts to change the dissolved salts to salts that do not form hard coatings and that can be easily flushed out of the system.

1B3f. *Hot water system*

The hot water system parallels the cold water system except that it uses a heat exchanger or a hot water generator to warm it and circulating pumps to keep it in motion. The hot water system generally is a loop system that returns the water to the generator and provides instant hot water at any outlet. The hot water generator is usually a steel tank lined with a noncorroding coating. It contains thermostatic controls that regulate the flow of the heating medium which usually consists of hot water or steam through copper tubes and copper baffle plates. The heated tubes and baffles heat the surrounding water usually to 140° and this water is constantly circulated. Hot water distribution piping is of copper or 85% red brass. It is softened and treated before it is heated.

1B3g. *Fixtures and accessories*

Most codes and specifications for office buildings or similar buildings now permit wall hung plumbing fixtures. Water closets and urinals are fastened to heavy yokes in the wall, and lavatories no longer rest on legs. This promotes cleanliness and sanitary conditions. All modern fixtures are of vitreous china which has a smooth glass-hard surface that can easily be kept clean. The architect engineer should be familiar with the constant improvements being made by the large manufacturers in the design of plumbing fixtures. Basins are made that can be fitted into marble or tile counters, color can be used in smaller installations, the faucets and drains are double chrome plated for easy cleaning, and to prevent corrosion, and so on. The

---

[4]pH is defined on page 179.

toilet room in any publicly used building is a practical necessity but there is no reason why it cannot be made attractive. The same criterion holds for soap dispensers, paper holders, towel disposal units, counters, and mirrors. For very little extra cost towel disposal units can be of stainless steel and set flush in the walls, the mirrors can be fitted with heavy chromium plated or stainless steel frames, the shelves can be stainless steel, and the paper holders can be doubled. All this gives the occupant the feeling of luxurious living at very little extra cost. It is paid for in the rent.

1B3h. *Plumbing tests*

Because the piping and devices that are installed under a plumbing contract all carry water, other liquids, or gasses, the requirements for testing the piping are strict. A summary of typical requirements for such testing follows: for water piping the system is closed and placed under water pressure that is at least 25 to 50% higher than the maximum working pressure will be. Such pressure can vary from 50 to 300 psig,[5] depending on the height of the building and the local code. For waste and vent lines, codes or specifications may require two tests. The first test for the rough plumbing calls for the lines to be filled with water which is kept in the system for several hours. Leaks can be detected by a falling water level. The second test takes place after all fixtures are connected and the water seal traps are in place. Pungent smoke or oil of peppermint are introduced into the system and leaks are detected by odor. Fuel gas lines are tested under air pressure with a pressure gauge to show any leakage. The pressure called for is usually 50% or more higher than the working pressure will be. Sprinkler lines are tested under high water or high air pressure depending on whether they are wet or dry. Standpipe systems are tested the same as sprinklers. All water-carrying systems also receive flow tests to measure the actual rate of flow in the lines against the specified rated flow. This is to make sure there are no obstructions or flow impeding bends in the lines. For obvious reasons, most of these tests are made before the system is closed in or covered over.

1C. THE DISTRIBUTION SYSTEM

The distribution system for a basic plumbing installation in a single-family residence, multiple dwelling, or office building consists of the same essential elements. It receives potable water from a private or public source and distributes it to sanitary facilities, wash basins, drinking fountains, air conditioning systems, food preparation areas, and to any other amenity requiring fresh clean water. It then receives the waste water from all these facilities and carries it to a private or public sewage system in such a way

[5]psig is pounds per square inch.

PLUMBING AND SPRINKLERS

as to protect the public from any effects adverse to health or convenience. A plumbing system also does other things such as disposing of storm water and helping to protect a structure against fire.

The residential system takes water from a well or street main by means of copper tubing (when under 3 inches), or cast iron piping. The water goes through a water meter and is then distributed by means of branch lines, usually copper tubing, to the heating boiler, to the hot water boiler, and to all cold water outlets. The hot water from the hot water boiler or generator parallels the cold water lines to all outlets. Good practice calls for individual valves on the supply lines to all fixtures and devices. The system then receives the wastes from all fixtures and after venting the lines to the outside air carries such wastes to a septic tank and drain field, private sewage disposal plant, or to a public sewage system. The piping in the waste system is either galvanized iron or cast iron. All fixtures must have water sealed traps and all lines to the septic tank or sewer must have cleanout plugs. Great care and strictly enforced codes keep the source of potable water and the sanitary disposal system completely and carefully separated.

The system obviously becomes more complicated as the structure gets higher and is used by more people. Without exception, such buildings receive their water from and dispose of their waste through public systems. In high structures, the water must be pumped up to higher floors, and to assure an even flow it is often pumped to a roof top storage tank. Such tanks have float switches which start a basement pump when the water falls below a certain level. Where there are no roof tanks the system is kept full by pressure-activated house pumps that run almost constantly. Sometimes a system is divided so that lower floors are fed by street pressure or there are separate pumps for various groups of floors. The actual distribution of hot and cold water is by means of wrought iron, copper, or galvanized risers running through centrally located shafts which are connected to branch lines on every floor. Each fixture on every floor is separately valved for both hot and cold water. In publicly used buildings such valves are controlled by keys to make them tamper-proof. The waste lines can be designed with offsets at intervals and with relief vent connections at every tenth floor or less to relieve gas pressure. The waste lines and vent lines are generally of cast iron except that lines under 2½ inches are of galvanized or wrought iron. Building codes specify minimum sizes of wastes and vents by assigning a fixture unit value to each connected fixture and relating the number of connected fixture units to such size.

The distribution systems for laboratory or medical buildings follow the same pattern insofar as sanitary facilities are concerned. The need in such buildings for liquids other than water and possibly for various gases makes them a special design problem. In some instances the sinks and waste lines must be acid-resistant. All codes prevent discharge of harmful laboratory or hospital wastes directly into a sewage system without prior treatment.

PLUMBING AND SPRINKLERS

1D. PROCESS WATER AND INDUSTRIAL WASTES

Process water can be described as the water that is used for industrial and maintenance purposes and which is kept strictly separated from water that is to be used for sanitary purposes, or for human or animal consumption. Without being treated, water can be taken from a river or other body of water and used for condenser cooling water in air conditioning systems, power plants, or for other industrial cooling purposes. At its intake it is usually screened to prevent large objects from entering the system and it is also treated with chemicals to prevent algae or other plant growth. The use of such public water for cooling has recently become controversial in many communities. Obviously, when the water is used for cooling it becomes warmer. The discharge of this warmer water in the huge quantities required for power plants is claimed to upset the ecological balance of the body of water (thermal pollution). It is said to promote the growth of undesirable marine plant life which uses oxygen from the water. The public outcry will force the architect engineer and the owner to spend time, effort, and money to provide means to cool such water down to its entering temperature before it is discharged. Huge cooling towers and holding basins have been mentioned.

The discharging of industrial wastes in general has become the subject for very strict code provisions and enforcement. The discharge of warm water into a public body of water may cause some damage, but the discharge of raw sewage or noxious industrial waste can be a very real health menace. The ever increasing pollution of our streams, rivers and lakes, and even our oceans, by unbridled or unthinking industry should be the object of concern to everyone, and especially to the architect engineer who can do something to prevent it. Many large city plumbing codes specifically mention wastes that are prohibited from entering the public sewer system. These include wastes that contain organic or inorganic solids that would deplete the dissolved oxygen supply of the body of water; noxious liquids or gases; liquids having too high an acid or alkaline content; radioactive wastes, and other wastes containing oils, greases, or like nuisances. All these wastes must be disposed of privately and an entirely new technique is now being evolved to study and control water pollution.

2. FIRE FIGHTING SYSTEMS

The fire fighting system for any large structure is an extremely important part of the design. Although it is installed by a plumbing contractor, it is a separate system from the regular building plumbing installation.

2A. STANDPIPES AND PUMPS

The standard way of supplying water for fire protection is by using a standpipe system. A standpipe system, usually located in a stairway, consists of a vertical pipe riser of black iron and of prescribed code size dependent on building height. The standpipe is connected to a permanent source of water and has hose connections on each floor. It is usually specified that the standpipe location and connected hose length are such as to reach within 30 feet of any part of the building. The permanent source of water can be a roof top water tank or a jockey[6] pump. The tank is used to supply domestic water and is kept full by a float-actuated house pump. When such tanks exist, the codes call for the location of the domestic supply outlet to be a distance above the bottom of the tank so that there is always a residual supply of water which comes out of the bottom of the tank for fire fighting. Codes allow either wet or dry standpipes, depending upon whether the pipe is exposed to cold weather. The wet standpipe is always kept full of water, either from the residual water in a house tank or by a jockey pump. This is a comparatively small electrically driven water pump actuated by a pressure gauge that shows when a standpipe has insufficient pressure. The jockey pump's purpose is to maintain the pressure of a wet standpipe. The dry standpipe substitutes compressed air for water. If a hose outlet valve is opened, the pressure drop immediately starts a pump or other water supply going to fill the standpipe system with water. All standpipes, whether supplied by a roof tank or a jockey pump, are connected to powerful fire pumps. Such pump specifications usually call for cast bronze impellers, heavy pressure seals, special bearings, and other corrosion and pressure resistant materials which must meet underwriter's and code requirements.

In addition to the requirements for roof tanks, or jockey pumps and fire pumps, all fire protection codes require an outside outlet from the standpipe, so that a fire department pumper can supply city water to the standpipe hoses. In cases of small fires the standpipe and the available building resources are usually adequate. Fire protection authorities, however, cannot take for granted that a standpipe is full of water or that a fire pump is in working order. The exterior outlet, called a *siamese connection,* consists of two 2½-inch fire pump hose connections. The locations of siamese connections are carefully prescribed. In case of fire, a Fire Department pumper connects two 2½-inch[7] hoses to the two pipes of the connection and is thus able to keep a 4- or 6-inch-diameter standpipe filled with water. All standpipe systems are subject to frequent inspection. Their efficiency and state of maintenance should be a matter of concern to owners and planners.

2B. SPRINKLER SYSTEMS

For positive fire protection in high risk areas the codes and/or the architect engineer specify sprinkler protection. A list of areas requiring sprinkler protection would include workshops where lumber, paint or high hazard

[6]This is explained below.
[7]This size is standard almost everywhere in the country.

flammable material is being processed or stored; below-grade storage areas; any below-grade activity; roof top cooling towers containing combustible fill; corridors and exit passageways in certain hazardous occupancy buildings, and in general any sensitive area where an incipient fire requires an instant response.

A sprinkler system can be described as an arrangement of piping with outlets so arranged that water may be discharged in a spray directly into the incipient fire. Sprinklers can be of the dry or wet type. The wet-type sprinkler, which always contains water under pressure, is fed by either storage tanks, fire pumps, or by street siamese connections through which a fire department pumper can deliver water. The dry-type sprinkler contains compressed air which on release through a sprinkler head actuates a valve to open the system to water, and simultaneously actuates a fire pump for supplying the water. The sprinkler heads are the water spray outlets. They are opened when the small metal plug that holds them closed is melted. Sprinkler head melting temperatures can range from 135° to 165° in normal occupancy spaces which are kept at normal room temperatures. In spaces where heat-generating work is being done, such as work involving retorts, kilns, steam cleaning, or baking ovens, such heads may be allowed to melt at temperatures up to 325°. A dry pipe system can also be a deluge system with open heads.

When sprinkler heads are actuated most specifications require that they cause a large alarm bell to ring to warn of a fire somewhere in the structure. Many buildings have their sprinkler alarm connected to a central station that warns a fire department. This illustrates the efficacy of a sprinkler system. It is a silent 24-hour-a-day watchman for fire safety.

A plumbing system is used for the very essential purpose of sustaining a healthful and pleasant environment, and is also used for such things as decorative fountains and other ornamental purposes. The architectural use of water for decorative purposes is a constantly growing field.

CONTENTS

1. DETERMINATION OF THE REQUIREMENTS
   A. Stairways
   B. Elevators
      1. Passenger
      2. Freight
      3. Worker and Material Transportation
   C. Moving Stairways
   D. Conveyors and Dumbwaiters
   E. Special Elevator Requirements in Very High and Multiple-Purpose Buildings

2. TYPES OF EQUIPMENT AVAILABLE AND TYPICAL SPECIFICATIONS
   A. Passenger Elevators
   B. Freight Elevators
   C. Safety Devices
   D. Door Operation
   E. Elevator Cabs, Openings, and Trim

11

VERTICAL TRANSPORTATION

1. DETERMINATION OF THE REQUIREMENTS

All structures over one story require vertical transportation. Stairways, elevators, and moving stairs supply the means for transporting people and materials. Without mechanically operated elevators, high-rise structures could not be planned and without hoists for material and workmen, they could not be built.

1A. Stairways

The stairway is the basic and only legal means of ingress and egress. When all else fails the stairway is always there. A tragic illustration of this occurred in a flash fire in a rather small building. Instead of running for the stairs the occupants tried to crowd into an elevator. Whether from overload or mechanical failure, the doors did not close. Eleven people died. They all could have been saved. Building codes are most explicit about the size and construction of stairways, their distance from the various parts of the structure, and the materials of the stair enclosures. The codes prescribe the allowable floor space per occupant for all types of structures and uses, and the maximum distance to a stair from the most remote part of these structures. The codes then specify the width of the stairways and stairway doors per occupant per floor. All stairs and stair enclosures in any structure over two

VERTICAL TRANSPORTATION

stories, except a private residence, must be fireproof and if the structure is over three stories it must have two separate means of exit. The architect engineer should familiarize himself with code requirements which may vary widely in different areas of the country. The codes, however, state only the allowable minimum and the architect engineer must program his individual building requirements. The future activity in the structure may require more stairways, or stairways with low risers and broad treads, or even ornamental stairways. These floor to floor transportation needs should be provided for as the structure is being planned.

1B. Elevators

1B1. *Passenger*

The mechanically operated elevator is both a product of the necessity for tall structures and the means by which they can be built and operated. It evolved when cities became crowded and structures had to rise vertically. The perfecting of the elevator so that it travels both speedily and safely has added immeasurably to the architect engineer's ability to design tall structures for all uses. The planner at the present time must provide for an elevator in any publicly occupied building of over two stories whether it be multiple dwelling, factory, or office building. He must determine what kind of elevator the occupancy requires, how long the occupant is willing to wait for an elevator, and from this and other data he can determine the speed, the size and the number of elevators he has to plan.

Elevator codes only specify safety requirements. The architect engineer and owner must decide on the size, speed, and capacity of the installation. They must bear in mind that along with good lighting and air conditioning, the speed and convenience with which the occupant is transported vertically can make or break a building. Waiting for an elevator is annoying and an intelligent planner tries to avoid such delays. In planning the passenger elevatoring for a building the architect engineer must have complete knowledge of the building's use and occupancy. There are tables in codes and other publications that set forth the normal occupancy in different structures such as: offices—100 square feet per occupant; apartment houses (habitable rooms[1])—140 square feet per occupant; retail sales areas—50 square feet per occupant, and so on. It must always be borne in mind, however, that these are minimum requirements and sometimes the occupancy may be much higher. The planner must carefully program the occupancy. An institutional building or a single-occupancy building may differ widely from a multiple-occupancy building with many small offices. The tenant of a single-occupancy building is able to control the time of the comings and goings of his em-

[1] Habitable rooms do not include kitchens or bathrooms.

ployees. In addition to occupancy standards, there are standards of acceptable time intervals between elevators. These vary from 20 to 30 seconds in an office building to 50 to 70 seconds in an apartment house. In smaller communities people are not as impatient as they are in a large city.

To start the calculation one assumes that the elevators should be able to transport a certain percentage of the total normal population of a building at peak traveling hours. Such hours occur at the start and finish of a work day and sometimes at lunch hour. For good service in a multioccupancy building it may be assumed, for instance, that a system should be able to transport 12½% of the occupants of an office building, or 7% of the occupants of a multifamily dwelling, in 5 minutes. To determine the handling capacity per elevator the first formula to be used is as follows:

$$\frac{60 \text{ (seconds)} \times 5 \text{ (minutes)} \times \text{number of passengers per trip (at peak time)}}{\text{round trip time (in seconds) (of elevators traveling at various rated speeds)}}$$

Now the architect engineer determines the number of passengers the elevators are likely to carry at peak times and the round trip time of the elevators in the banks traveling at various speeds. This is difficult but not as difficult as it sounds. Elevator engineers through long experience have determined certain standards, and the experienced architect engineer should be able to make certain assumptions and check their figures. The number of passengers per trip may be assumed to be about 80% of the carrying capacity of a car. Let us take as an example the passenger car specification for a 30-story office building that will contain 18,000 rentable square feet per floor and is therefore assumed to be occupied by 180 people per floor. For good service it is customary to have a bank of elevators serve a maximum of 12 floors and fewer, if possible, because on occasion an elevator may make all 12 stops.[2] In this building there are three banks of elevators each serving ten floors. In the top bank of a 30-story building where the first stop is at the 20th floor, the architect engineer might start with a 3500-pound capacity car traveling at 1000 feet per minute. The car capacity and speed is chosen because it is the largest economical size for that speed. A car of 3500-pound capacity will normally carry 19 passengers. Stopwatch studies of passenger loading have shown that a car of this capacity and speed will make a round trip at peak uptravel in about 140 seconds. (There are available tables or charts that show these figures.) This means that 19 passengers are carried every 140 seconds or 40 passengers in five minutes. If 12½% of the 1800 occupants of the top ten floors are to be transported in five minutes

---

[2]This happened once in a very fine office building when the head of a large firm on the ninth floor was on his way up to his office. After the car had made nine stops the tenant stormed in to his office, called the owner and threatened to break his lease unless something was done about it. Something *was* done—the elevator starter was instructed to "spot" one elevator when this executive appeared. Spotting will be explained later.

VERTICAL TRANSPORTATION

then six cars will be required. The waiting interval is calculated by dividing the 140-second round trip time by six cars, which comes to 23 seconds. This is exactly what a specification for a good high-rise building should call for, namely: high-rise bank—floors 20 to 30—cars 1 to 6—Capacity 3500 pounds —Speed 1000 fpm. Such a specification provides the building with much flexibility. There is enough leeway in the time interval and the carrying capacity to enable the owner to shift floor occupancies, or to rent space to a tenant who might wish to use his space for high-density classrooms, training, or conference purposes. In a case like this, it is a great temptation to specify five instead of six cars. The time interval would be 28 seconds and only 11% of the population could be transported in five minutes. The hitch is that this marginal service will do until a car is taken out of service or a tenant crowds his space by changing its use. It is done in marginal buildings but it is not recommended. Once a building elevatoring system is built, it is extremely difficult and very expensive to add to it. The elevator systems of other than high-rise office buildings must also be given careful attention but the requirements are not as onerous. In a three- or four-story suburban office building with 20,000 rentable square feet, a single elevator of 2000-pound capacity at 200 feet per minute will, at peak time, according to the tables of standards, carry 11 people per round trip of less than 100 seconds duration. This means that in 5 minutes it will carry 33 people or 16% of the population of 200 people. Of course, anyone just missing the elevator will have to wait 100

$$\frac{60 \times 5 \times 11}{100} = 33$$

seconds. By spending slightly more money the elevator can be speeded up to 350 fpm and the waiting interval reduced. This is only at peak times. The traveling time will be much reduced when the elevator cab is not filled to capacity with the resultant delay in loading and unloading. For apartment houses, hotels or other structures, the comparative need for people to get where they are going and how quickly they must get there must be programmed. Elevators, especially fast ones, are expensive. Normally, people who do not have an urgent need to get to a desk or an appointment will wait with reasonable patience. The architect engineer should be aware of this and take advantage of it.

1B2. *Freight*

The requirement for freight elevators obviously varies greatly with the activity to be carried on in the building. The normal use for a freight car in an office building is for carrying supplies to offices, transporting coffee carts, cleaning personnel, materials, and carrying rubbish. It may occasionally be used for transporting a heavy or bulky load. In an office building where

VERTICAL TRANSPORTATION

normal office activities occur, a single freight elevator for a building of 600,000 square feet and 30 stories height is usually sufficient. It will serve every floor and it should be of a larger capacity than the passenger cars and should also have a much higher head room for bulky loads. The door openings should be as wide as possible. It can be slower in speed and a typical specification would call for a 4000-pound car travelling at 500 fpm It is well also to equip at least one passenger car in each bank with protective blankets so that they can be used when large quantities of material have to be moved quickly. In one building the author knows a single tenant occupying many floors had so much catalogue and other mailing activity that a single freight car was not enough. A passenger elevator shaft had to be brought down to a basement floor adjacent to a truck loading dock. It was expensive, and removing the car from service caused great inconvenience. It might have been possible to foresee this, had a very careful study been made of the tenant's activities. In the small office building or apartment house the freight elevator will simply be the passenger car or one of the passenger cars with protective blankets. Such combination elevators also occur in hospitals where a car must be large enough to handle a stretcher table.

In industrial plants where heavy loads are carried by lift truck or mechanical hand trucks, the freight elevator must be very sturdily built with wide opening doors and high head room. Such general purpose elevators will run up to a 10,000-pound capacity with a speed of 50–100 fpm. Such a car may measure 8′ x 12′ and have an 8-foot-wide door. Special duty freight cars can be manufactured to carry fully loaded trailer trucks or even fully loaded railroad freight cars. There are freight cars made for light duty that can be installed in a hoistway without having to reinforce the roof for overhead sheaves or the building to support the rails. All the machinery is in the basement. It can almost be called a packaged unit. The architect engineer will find that elevator manufacturers will provide any requirement he may have for transporting anything vertically.

Another version of the freight car is the sidewalk elevator. This elevator can serve one or two basement floors from an opening in the sidewalk. The familiar version is the bow topped elevator platform which pushes up a pair of sidewalk doors. The machinery is in the basement. Such sidewalk elevators are rated for 2500 pounds and travel quite slowly. Safety codes require the presence of a person on the sidewalk and a loud ringing bell to warn passersby from stepping on the doors when they are lifting. Most city codes now forbid their use because any doors opening on a public sidewalk are inherently dangerous.

1B3. *Worker and Material Transportation*

It was stated earlier in this chapter that without mechanically driven vertical transportation our modern high-rise building could not have been

VERTICAL TRANSPORTATION

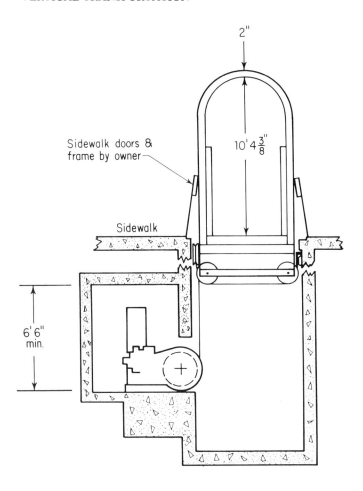

Sidewalk Elevator.

built. Vertical transportation for construction materials preceded the vertical transportation of men. Derricks have lifted steel and other heavy materials, and so-called hod hoists have lifted plaster and brick and lumber for many years. These machines, however, have no safety devices and safety codes prohibit their use. It was the custom, and still is, as the building progresses upward, for the elevator people to install a temporary elevator machine to take men part way up the building. However, because of its semipermanent nature such temporary elevators have to wait for shafts and overhead steel to be completed. In recent years several manufacturers have perfected dual-purpose man and material lifts. These can keep pace with the upward progress of the building in their own shafts, which are built of pipe steel, and are accepted by city and safety authorities. While the use of such devices is more the builder's than the architect engineer's responsibility, the architect engineer should be aware of the possibility for saving time and money, especially on a cost-plus contract.

1C. Moving Stairways

The electric stairway or "escalator" is a means of vertical transportation by which the planner can move a large mass of people and keep them moving from floor to floor. The continuous movement of this stair is attractive to people because they can use it without waiting. It is used in office buildings where the owner wishes to attract large numbers of people from his street floor entrances to upper or lower floors. The presence of a bank, or brokerage house or restaurant on a mezzanine or second floor, or a restaurant or display area below street level, makes such a system desirable. Typical examples of the use of moving stairways in large office buildings can be cited from the author's own experience. In one building, there is an underground entrance from public transportation in addition to the street entrance. The large employee cafeteria is on the second floor. Both these facilities require the vertical movement of large numbers of people to or from the street floor (first floor), from which all the elevators start. In the case of the cafeteria, studies showed that the best and most orderly service to employees would be provided if they could arrive at the food counters in a continuous stream rather than in abrupt large groups. This was accomplished by carrying them all to the first floor by elevator and having them separate on a moving stair to the cafeteria floor. The below ground entrance, especially in bad weather, needs an escalator to move people to the first floor elevator lobby. In another building, the plans called for a large front plaza. In bad weather, people leaving public transportation would have to make their way across this plaza without shelter. The problem was solved by providing a kiosk to cover a pair of moving stairways leading to an underground passage, and another pair to bring the people up to the first floor of the building. This underground passage was lined with shops whose rent more than covered the extra cost of providing this service. The use of the moving stairway in department stores, terminal buildings, display buildings, etc. is a means whereby a skillful architect engineer can direct his people to where he wants them to go. The most attractive displays are placed where people can see them from a moving stair. The entrances and exits from trains, buses, and planes are situated so a moving stair will transport people to or from such spots, quickly and without delay.

Moving stairways are manufactured in two standard widths and in two standard speeds. The 32-inch-wide stair allows for an adult and a child to stand side by side while a 48-inch-wide stair allows for two adults or for one to pass another. At maximum capacity, a 32-inch stair traveling at 90 fpm can carry 425 people in 5 minutes or just over 5000 people per hour. The 48-inch-wide stair can carry 680 people in 5 minutes or 8000 per hour. At 120 fpm, to which either width stair can be adjusted, they can carry from 6700 to 10,700 people per hour. These are large numbers and the planner should make use of them wherever possible.

VERTICAL TRANSPORTATION

The moving stair can be installed in a criss-cross or parallel pattern, or may be adjacent or separated. The architect engineer and owner can choose a system based on the room available and on where they desire the stairways to carry people. The criss-cross, where the up and down stairways form an X, is the most widely used and provides the best circulation because it separates people going in different directions. The parallel up and down stairs are more decorative but not as efficient. The other patterns are used where economic and other circumstances warrant. The architect engineer must always bear in mind that the huge people-carrying capacity of moving stairs places the burden on him to make full and efficient use of this medium.

1D. Conveyors and Dumbwaiters

As a supplement to the elevator, many buildings, especially those occupied by a single enterprise, have included a conveyor or an electrically operated dumbwaiter. These devices carry papers, mail, and small parcels

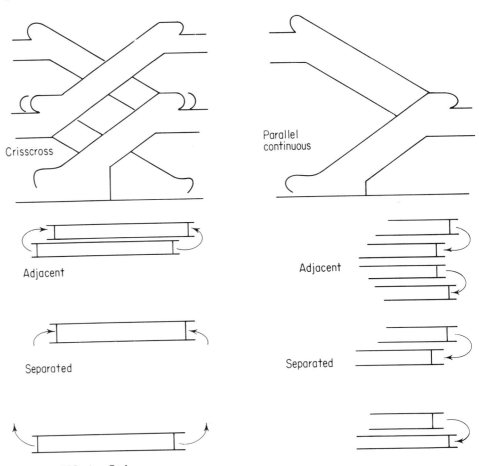

Patterns of Moving Stairs.

VERTICAL TRANSPORTATION

Portion of an Electric Stairway.

throughout a building without the use of messengers, and require only a single clerk or less per station to handle incoming and outgoing material. In the past few years the Post Office Department has been urging all buildings to install a vertical mail conveyor in order to avoid the use of manpower in transporting mail up through the building. The usual conveyor consists of a pair of endless chain belts traveling over power driven sprocket wheels. Platforms or open cars are fastened to the chain at intervals and these platforms carry mail trays, which contain a device that can be set to push these trays off the moving platforms at their destined floors. Such trays can carry 40 pounds. Conveyors can be built to transfer horizontally as well as vertically. Dumbwaiters are simply small elevator cabs built to carry all sorts of material between floors. A dumbwaiter installation can carry up to 500 pounds and travel up to 200 feet per minute. The dumbwaiter can carry more in a single load but the conveyor is more flexible and is completely automatic. The planner must be completely familiar with the client's requirements before specifying either a dumbwaiter or conveyor.

1E. Special Elevator Requirements in Very High and Multiple-Purpose Buildings

When an architect engineer plans a building, along with providing maximum comfort and convenience to the occupants, he must provide the owner with maximum rentable space. He must supply a service core that contains air ducts, electric shafts, water lines, toilet rooms, stairways, and elevators. When a building gets up above 60 stories, it requires so many elevators to carry the tenants expeditiously, even at high speeds (1800 fpm or more), that the lower part of the building is in danger of losing a very large portion of its interior rentable area to elevators. To overcome this the planners

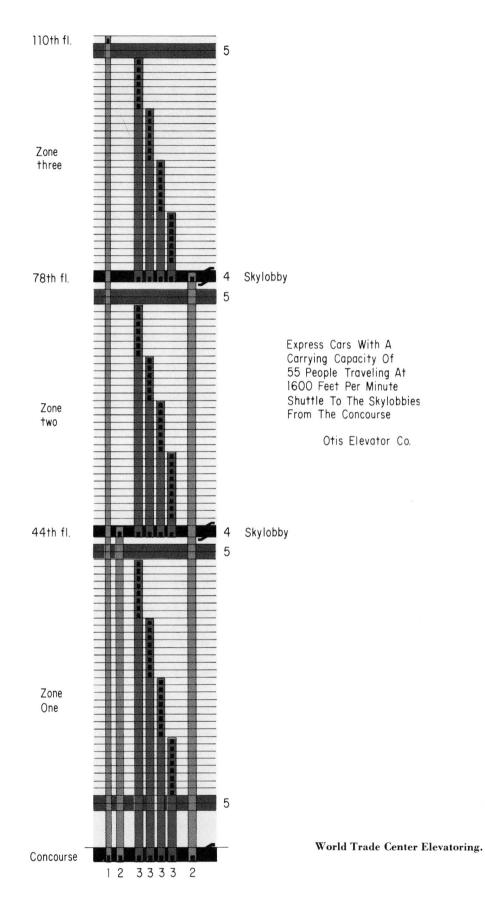

VERTICAL TRANSPORTATION

have come up with a new idea which might be called the *sky lobby*. This lobby, situated high in the building, serves as a transfer point from one set of elevators to another so that the architect engineer can literally pile one elevator system over another. In the John Hancock Building in Chicago, which contains 44 stories of apartments over 40 stories of offices, there are two elevator systems with the upper apartment system being served by a separate street lobby as well as a "sky lobby." High-speed large-capacity (4000 pounds at 1400 fpm) shuttle elevators carry the apartment tenants 44 nonstop stories to their own elevator system. Typically, the sky lobby contains attractive shops and other diversions. This is not only profitable but tends to make the tenant more patient when having to make an intermediate transfer stop. The same situation occurs in the new World Trade Center buildings in New York City. Here three elevator systems are piled one on top of the other. The bottom system serves floors 1–44, the intermediate serves 44–78 and the top 78–112. In this case one set of shuttle cars (10,000-pound capacity at 1600 fpm) rises 44 stories nonstop to the first "sky lobby" and another set of shuttles rises 78 stories to the second one. It would be too much to ask upper floor tenants to transfer twice. There are a total of 98 passenger elevators serving all the floors and if they all had to start at the street floor there would not be much space left to rent.

There are other applications of elevators that are more spectacular than useful, such as the outside elevators in several hotels. These have to be supported on one side only and require very careful engineering and special safety devices.

2. TYPES OF EQUIPMENT AVAILABLE AND TYPICAL SPECIFICATIONS

As the architect engineer and owner are deciding on the type of elevatoring that the building requires they should become familiar with the types of available equipment that are adaptable to their particular needs. A complete knowledge of this subject will enable the architect engineer to select the most economical installation to serve his calculated criterion of capacity, waiting interval, and general convenience to the occupants.

2A. PASSENGER ELEVATORS

The variety of passenger elevators available as to speed, capacity, and operating characteristics is such that a stock-built model can fit almost any reasonable requirement. In a low office building or apartment house of ten or twelve stories the geared traction machine can be used for speeds up to 350 fpm. The geared traction machine consists of a high-speed electric motor that drives the hoisting sheaves by means of a worm and gear reduc-

VERTICAL TRANSPORTATION

tion unit. The motor may be AC, in which case the electric current is taken directly from the power lines, and the speed of the motor is controlled by a rheostat and the application of a braking action on the car rails. This method of leveling the car is inexact and can and does cause a tripping hazard. The operation of the car is also likely to be jerky. It is economical but not recommended. By varying the ratio between worm, gear, and drive sheave, this machine gives a wide variety of speeds and capacities. It is better to employ a DC motor to take full advantage of the geared traction machine. The DC motor receives its operating electric current from an AC–DC motor generator set. The variable voltage that can be applied to the DC motor from the generator brings utmost flexibility in speed, and enables the operating motor to take up its load evenly, and bring the car to a smooth stop.

Westinghouse Geared Machine
1 Grooved traction sheave
2 Worm and gear reduction drive
3 Helix angle adjustment
4 Brake assembly
5 Flange mounted motor

Geared Traction Machine.

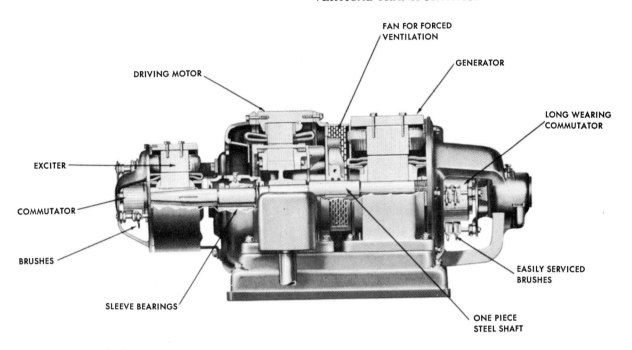

Motor Generator Set.

The signal and control system for such a slow moving machine can be operated by the use of "selective collective control," which is sufficiently flexible to handle the moderate traffic in a small building with one or two elevators. The car answers calls in the direction in which it is traveling regardless of the sequence in which the calls are made. The car will reverse its travel at the highest or lowest corridor call. Unanswered calls registered in the opposite direction will be answered on a subsequent trip. The pattern can be changed to permit the car to reverse its direction at the highest registered call, following which it will return to the lobby and remain there with its doors open.

The passenger elevator meets its real challenge in the modern high-rise office building where thousands of people arrive and depart and go to lunch in very short periods of time. The operating mechanism in such cases is always a gearless traction machine. This consists of an AC–DC motor generator which drives the DC hoisting motor. This DC motor is directly connected to a sheave over which the hoisting cables travel to raise and lower the elevator and the counterweight. Such gearless traction machines can operate elevators of 10,000-pound capacity, traveling at 1800 fpm. A specification for a typical high-quality office building would contain the following: The gearless motor shall be designed to meet the severe conditions of elevator operation and shall consist of a slow speed direct current motor directly connected to a traction sheave and brake wheel. Bearings shall be rigidly mounted in proper alignment and shall be of ample capacity to sustain the heavy loads supported by the machine. Sheave and brake wheel shall be machined from semisteel

VERTICAL TRANSPORTATION

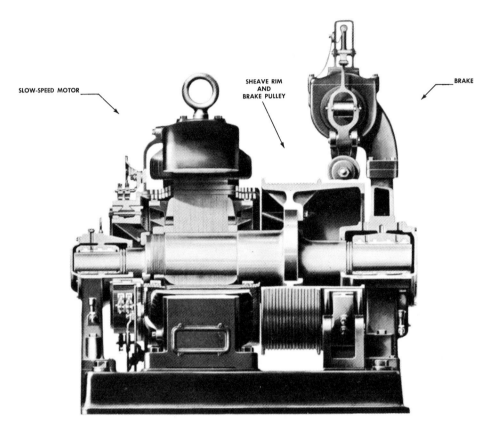

Gearless Traction Machine.

castings. Grooves in sheave and surface of brake wheel shall be carefully machined and both pressed and keyed on a heavy forged steel shaft. The brake shall be actuated by electrically released[3] heavy springs and shall be suitably lined. The control system shall include a motor generator set which shall apply uniformly varying direct current voltage to the hoisting motor. The operating system shall consist of a series of push buttons in the car corresponding to the landings, all connected to a dispatching system and control switches governing floor selection, up and down travel, acceleration, and retardation. After a car has received a start signal, and the doors have closed, and the safety interlock circuit has been established, the car starts in the direction established. It then stops in sequence at each floor for which a car button or corridor button has been pressed, regardless of the order in which the signals registered. Provision is made to eliminate any car from the system if it fails to start at a predetermined time and every car is weighed so that a fully loaded car bypasses all corridor calls.

[3]The braking system on an elevator is a "fail safe" system. When the current is off the brake goes on.

The sophisticated control and signal system, responding to computer demands, can change the traffic pattern in seven ways: the morning inrush; balanced two-way traffic; heavier up; heavier down; heavy down; evening outrush, and off-hour. This permits the elevator car to be at its most-needed location with the smallest possible delay. As an example, for the morning inrush the system will start lobby cars upward at very close intervals and as soon as a car starts it will immediately establish a demand at the lobby to provide one loading car and two standby cars for "up" service. It reverses the direction of a car downward as soon as its last passenger has been discharged. Such constant service can empty a lobby very quickly. The other patterns have been established through constant observation of passenger travel in an office building and are calculated to render the maximum convenience in transportation.

2B. Freight Elevators

The freight car in an office building is a car capable of doing all types of work. It carries workmen, materials, and miscellaneous heavy loads all day long and is put to much harder work than a passenger car. The freight car is always run by a geared traction machine, and unlike the sophisticated passenger cars, has the simple "selective collective control" system that answers "up" calls on the way up and vice versa. A freight car in a large building, in addition to being automatic, is almost always run by an attendant during business hours and therefore is frequently equipped with a drop annunciator which visibly registers the floor that has called. It is relatively slow moving at 500 fpm in most cases, and has a large capacity and high head room.

2C. Safety Devices

The safety devices on an elevator are both positive and "fail safe." To start with, no car can start moving until both the car door and the shaftway door are closed and have thereby closed the operating circuit. This is called the door *interlock*. Elevator hoisting cables are of special steel and are especially woven to take the constant bending and stretching, and are built to standards as specified in the American Standards Association Code. They are usually 6×19 or 8×19 which means that 19 wires are wrapped first to form a strand, and then six or eight strands are wrapped to make the cable. The cables are double wrapped around the hoisting sheave and a secondary idler sheave, as shown, to provide for any possible slippage. The counterweight, which is usually equal to the weight of the car plus 40% of

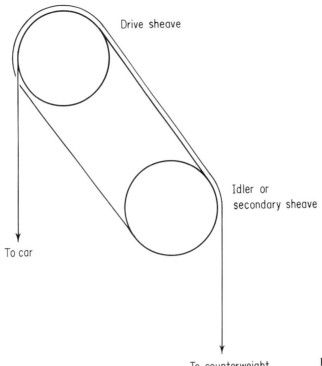

Double Wrapping of Cables.

the car capacity, not only makes the work of the hoisting motor easier but provides a certain safety factor. If the car were loaded to less than 40% of its capacity the counterweight would pull it up in the event of a failure of some kind. The rails on which the car and counterweight ride are of special machined steel and manufactured in accordance with ASA Code. A limit switch is provided at each terminal point, its function being to cut off the electric current if a car should override the final landing. Should the car start to travel faster than its preset speed, a governor cable which travels at the same speed as the car will trip a governor and open a switch cutting off the power. The shut down of the power automatically applies the brakes to the brake drum. Should the car, however, continue to gain speed, the governor will physically clutch the governor cable and trip a powerful set of wedges which will clutch the rails, preventing any further down movement. These wedges can only be released by the upward movement of the car. As an additional precaution, all elevator shafts must be extended into an elevator pit below their bottom landing and the pit must be fitted with bumpers whose motion can be dampened either hydraulically or by means of heavy springs.

2D. Door Operation

The smoothness and speed of elevator service depends to a great extent on the operation of the car and shaftway doors. The elevator cannot run until the doors are closed, and therefore the speed and convenience

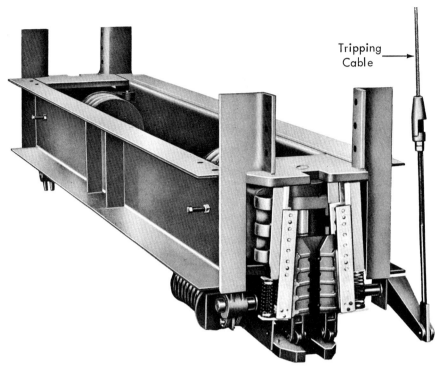

Westinghouse Rail Safety Clamp.

of the elevator operation depends on the efficiency of this opening and closing operation. Stopwatch timing of door opening and closing shows an average of 7 seconds for good operation. This multiplied by the number of possible stops of a car further emphasizes the need for efficient devices, especially in the high-speed service that has become a necessity for the modern high-rise office building. The architect engineer recognizes this in his specification which will mention the following items: Hoistway entrances shall consist of structural steel angle struts securely fastened to the building structure and connected to steel door hanger supports. The struts and hanger support shall be covered by 14 gauge steel fascia plates and there shall be an extruded aluminum sill securely fastened to the frame and the building structure. The specification up to this point has called for a strongly built securely fastened, and rigid door opening to provide an unyielding support for the door operating mechanism. The doors themselves are specified to be 1¼-inch thick of 14 gauge furniture steel substantially reinforced and soundproofed. The mechanism by which the doors are actually operated consists of a DC motor which through a gear reduction unit drives both the car door and shaftway door to an open or closed position. The actual motion is imparted to the doors through positive linkage between the door and the gear mechanism which causes the shaftway door to come to an open or closed position. The positive linkage means that rigid steel arms perform the work rather than chains, belts, or cables, which can develop play and throw off timing. The

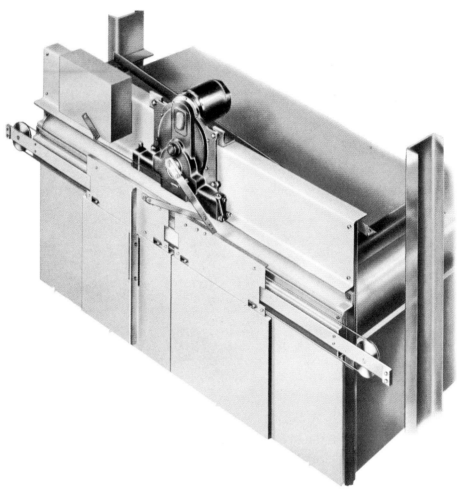

Otis Electric High-Speed Door Operator for Passenger Elevators.

doors are supported on the rigid door hanger support by sheave wheels made of sound-reducing material which rotate on precision roller bearings. To provide further positive support there is a roller on the underside of the hanger track to hold the door closely to the track. As the doors open or close, a hydraulic cushioning device reduces the motor torque and brings the doors to a smooth stop. The shaftway doors are actuated by the same motor mechanism as the car doors, and open and close simultaneously with the car doors. This is done by means of a projecting vane on the car door that engages the rollers on the shaftway doors. When the car doors close the vane disengages itself and the closing of both sets of doors actuates the interlock, which by closing an electric circuit allows the elevator to proceed. All doors have a safety mechanism attached to them which prevents a door from closing when a person or object is in the door opening. This consists of a leading edge on the door which if pressed reverses the motion of the door and causes

VERTICAL TRANSPORTATION

it to open again. Some systems also have an electric eye, which if interrupted does the same thing. No elevator can start until both shaft and car doors are closed. As an added precaution, the shaftway doors are counterweighted to ensure positive closing. Every part of the mechanism is built for speed, smoothness of operation, and safety.

2E. ELEVATOR CABS, OPENINGS, AND TRIM

Much of the convenience, speed, and safety of the elevator service can go unnoticed by the traveling public if the door and cab finishes are of poor quality or design. The public assumes that elevators are safe but it sees only the cab, the doors, the push buttons, and the hall lanterns. If these visible objects make a good impression, the public will assume the other components to be of equal quality. The architect engineer first specifies sturdy construction. Thus the cab walls may be of 16 gauge furniture steel covered with plywood on which the final finish is mounted. The doors may be of heavy-gauge furniture steel[4] with a carefully specified baked enamel finish, or of anodized aluminum, stainless steel, or even brass or bronze. The cabs should be provided with handrails, and ceiling ventilation, (and music in some cases). The car illumination should be decorative with luminous ceiling panels or softly colored lighting. The floors should be of pleasing vinyl tile or carpeting. Depending on use, the wall decoration may be of wood paneling, vinyl, stainless steel woven mesh, or any other decorative material. It should be attached so that it can be easily removed and replaced if it is damaged (and it will be!).

The push buttons should be conveniently located and tamper proof. They should indicate that they have been pushed by means of a light that stays on until a car answers the call. The car signal lanterns should be attractively designed and so located that they may be easily seen from all parts of the elevator lobby. They should be of two different colors for "up" or "down." There should be an audible signal device attached to the lanterns to indicate that a car is about to stop.

Since a visitor's first impression of a building is formed by the main entrance lobby and the elevators, it follows that elevator design is of extreme importance in the building plan.

[4]Furniture steel is specially flat rolled steel with no waves, bends, or imperfections in its surface.

CONTENTS

1. THE DETERMINATION OF THE DESIGN
    A. The Masonry Wall
    B. The Structural Concrete Wall
    C. The Structural Steel Wall
    D. Prefabricated Industrial Walls
    E. Curtain Walls
        1. Some examples of designed curtain walls
            a. A typical specification for aluminum and glass
            b. Description of masonry and glass wall and specification
        2. Glazing practice
            a. A typical specification
        3. Prefabricated curtain walls (pivot windows)

2. MOCKUPS, PERFORMANCE STANDARDS, AND TESTS

3. SOME EXAMPLES OF SPECIALLY DESIGNED WALLS

12

THE EXTERIOR WALL

1. THE DETERMINATION OF THE DESIGN

The architect engineer, before and during the design of a structure, must constantly bear in mind the intended use of the structure. He must design his structural frame and floors to carry certain loads. The air conditioning, lighting and plumbing must meet the needs of the occupants of the building and their work activity. The module to which the lighting, partition system and column centers are designed is very carefully calculated to be the most convenient for desk spacing and office sizes. The size and shape of the structure have been decided by economic requirements, zoning, or both. It is not until the exterior of the building is designed that the architect engineer can devote his energies to aesthetics. He can clothe the exterior with glass, masonry, aluminum and glass, bronze, or any other suitable material. The material chosen must be able to withstand the prevailing weather without disintegrating or causing severe maintenance problems. It must be able to withstand heavy winds, rains, or snow without leaking air or water. It must have sound and heat insulating qualities. The exterior of a building should express the desire of the owner and architect engineer for quality, institutional prestige, and monumentality. For instance, in a small apartment building a feeling of quiet and intimacy is desirable. This expression of human emotion in architecture is very difficult. All too often the exterior is left to chance and frequently shows it!

THE EXTERIOR WALL

1A. The Masonry Wall

The use of masonry for exterior walls is gradually giving way to the machine-made factory-assembled metal and glass walls, especially in small office buildings, but masonry is still extensively used in small construction. The garden apartment, medical or research building set in a special zone in a suburban community is very apt to have a masonry exterior. School buildings and college dormitories have been designed for masonry by famous architects. Masonry is sturdy and weatherproof and in skillful hands can present pleasing and even startling effects. The most common form of masonry wall is the brick wall. This is sometimes laid as a solid wall from 8 to 16 inches thick, depending on the weight it has to carry. Building codes specify the thickness of walls depending on the height of the building but in cases where masonry walls carry the weight of the structure (wall bearing), the architect engineer must know the compressive strength of the material he plans to use. There are ASTM standards for clay, shale brick, for sand lime brick, and for concrete brick. The clay, or shale brick is the strongest, the most weather resistant, and most decorative. It comes in many colors and textures and can be laid in various patterns. The mortar used in masonry

Stiles and Morse Colleges, Yale University. Eero Saarinen and Associates, Architects. Courtesy Ezra Stoller Associates, Inc.

THE EXTERIOR WALL

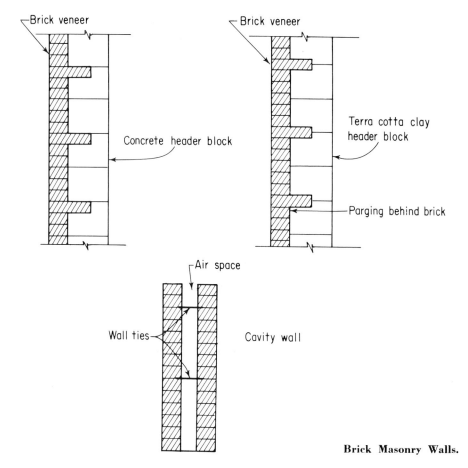

Brick Masonry Walls.

construction is also the subject of codes and ASTM standards. It usually consists of Portland cement with a small proportion of hydrated lime or lime putty to make it workable, and sharp clean sand which should be in the proportion of not less than 2 to 1 and not more than 3 to 1 to the Portland cement and lime mixture. The brick exterior wall can be built of various combinations of brick and hard burned terra cotta, or clay hollow tile, or concrete block. In any case the face brick must be bonded to its backing at every sixth course as required by most codes. The architect engineer, in specifying the laying of a brick wall or any masonry wall, should realize that the numerous mortar joints may shrink or crack and the brick itself may not be entirely impervious to moisture. He specifies that the brick be laid in full bed joints and have full cross joints. Very often there is a waterproofing compound added to the mortar. Most specifications call for "parging" or covering the interior face of the brick with a coating of mortar as a protection against moisture. The "cavity wall" makes the structure impervious to moisture penetration and by enclosing a dead air space between two independent brick or block walls it also insulates against heat loss. The two independent walls are built two to three inches apart and are bonded by metal ties.

THE EXTERIOR WALL

The air space must be kept clear during and after construction. Codes are specific about the support of masonry walls. They usually state that a wall must be laterally supported at intervals not exceeding 20 times the thickness of the wall. This means that for an 8-inch wall there must be a lateral support every 13 feet either horizontally or vertically. The brick or masonry wall can be dressed by the use of glass brick, cut stone trim, or glazed terracotta. Each of these materials must be used within its defined strength and must be carefully bonded to the surrounding masonry.

1B. The Structural Concrete Wall

In Chapter 6 under "Structural Concrete" (Section 2D) it was stated that in many structures built of reinforced concrete the exterior walls or portions of them are poured with the structure. Because of the plasticity of concrete the architect engineer can achieve bold and interesting shapes and surfaces. The famous architect, Corbusier, has designed entire cities of concrete buildings and whether aesthetically pleasing or not they are certainly imaginative and strike a new note in the use of this material. With the advent of the computer it is now practicable to perform the necessary and

**Garage in New Haven, Connecticut. Paul Rudolph, Architect.
Courtesy Ezra Stoller Associates, Inc.**

THE EXTERIOR WALL

Ronchamp Chapel, France. Corbusier, Architect. Courtesy Ezra Stoller Associates, Inc.

intricate calculations when a concrete outer wall becomes a part of the structure. In a conventional structure such outer walls can be confined to columns and exterior spandrel beams to which metal framed windows can be added. The concrete itself can be cast with specially colored aggregate which can be exposed by acid etching or steam cleaning after the forms have been removed and before the concrete attains its final hardness and strength. When an exterior concrete wall is cast in place as an integral portion of a reinforced concrete structure the architect engineer pays particular attention to it because this is what the public will see. The quality of the building will be judged by the appearance and workmanship of this wall. The mix of the wall is carefully specified by the proportion, the color, the size of the fine and coarse aggregates, and the type of cement to be used. Samples are called for before the final mix is decided. The specification details the way in which the various materials are to be weighed and mixed at the "batching plant." The reinforcing steel must be placed with great care and as far from an exposed surface as possible.

To avoid any appearance of joints on a concrete face, the forms must be lined with a continuous membrane or an applied form coating. The forms themselves must be very strongly and carefully built to be able to hold the heavy mass of wet concrete without deflection. The concrete for exposed architectural concrete is poured as a stiff mixture with only a 2-inch slump allowable in a 12-inch-high cone.

THE EXTERIOR WALL

The specification for placing the concrete contains the following precautionary directions:

1. The concrete must be transported from the mixer to the final locale as rapidly as possible and held together to avoid a separation of ingredients.
2. The concrete must not be allowed to fall more than four feet.
3. It must be placed as closely as possible to its final position to avoid rehandling.
4. It should be placed in horizontal layers not deeper than 24 inches.
5. The concrete must be carefully placed around the reinforcement to avoid displacement of steel.
6. It must be carefully and thoroughly vibrated to form a compact mass.
7. When the coarse aggregate is to be exposed, the concrete must be placed in vertical layers and the vibrator run so the coarse aggregate is pushed against the form before it is followed up with the finer aggregate.
8. When the concrete is to be sandblasted the forms must stay in place at least 30 hours for walls and 48 hours for beams. This still leaves the concrete soft enough to accept sand blasting but set enough to avoid collapse. This removal of forms in such a short time must be done very carefully and positive means must be taken (shoring) to avoid any possibility of failure.
9. Finally the concrete must be thoroughly cleaned and then given at least two coats of moisture sealer.

The entire procedure, if followed carefully, will produce an evenly colored, evenly textured, crack-free building wall.

1C. THE STRUCTURAL STEEL WALL

In recent years the major steel companies have developed steels that when exposed to the weather form a protective compound of oxides and after a few years of aging turn a very attractive dark russet color. This coating is not only maintenance free but protects the steel beneath it from any further corrosion. The architect engineer is now free to design a structure which uses working steel for the exterior wall. In addition to the exposed columns he can use plate girder spandrels and between them cover the entire exterior wall surface with steel, leaving only the space between the girders for windows. In another expression of steel as an exterior wall the World Trade Center buildings in New York City are using closely spaced structural steel columns tied together by massive horizontal spandrel beams as a bearing wall which in addition provides the wind bracing necessary for a building of this height (110 stories).

The building exterior walls are of such closely connected steel, both vertical and horizontal, that they are really square box beams in which the only openings are the narrow ones between the exterior columns. Although in this case the steel is covered with aluminum sheets it can still be considered an expression of the steel structure rather than a curtain wall. Because of the enormous wind pressures that will be encountered at these heights the designers have taken elaborate precautions to prevent leakage of wind-driven rain through the exterior wall. This has been done by providing a void space between the outer aluminum and the inner steel. This space, which is opened to the outer air acts to equalize the wind pressure and prevents moisture from driving through. There is constant research in the use of working steel as an exterior wall. The advantages are many.

1D. Prefabricated Industrial Walls

No discussion of exterior wall material can be complete without mention of the many varieties of prefabricated exterior wall panels that can be used for industrial buildings, warehouses, auditoriums, and other buildings where long walls without windows are required. The architect engineer has a wide choice in color, texture, weather resistance, material, and general architectural quality. For larger structures he can also use most of these materials cut to required size in designed curtain walls. Such a complete solid wall without openings offers a distinct economical advantage. Here the cost of erection labor is at a minimum and the colors and textures of these large wall panels is designed for such use. All of these wall panels are of laminated construction (built of several layers of material), which combine sound and heat insulation and weather resistance within the panel. To describe a few available materials: A steel sheet with corrugations for strength and coated with zinc to protect against corrosion, then further covered with asphalt-impregnated asbestos felt, and finally covered with a coat of hot waterproof paint; a sandwich of two aluminum sheets with insulating material as a core; an aluminum exterior sheet with a hardboard or plywood interior sheet with a cellular polystyrene core and an exterior coated with baked synthetic resin or vinyl. There are many other combinations of materials and shapes, and each one has qualities that may be uniquely fitted to the building's requirements. The architect engineer should be familiar with all the choices before making a decision.

1E. Curtain Walls

The curtain wall in modern construction is an expression of the assembly line technique in architecture. It replaces the "hand made" exterior, built brick by brick with large prefabricated sections that are factory produced. In many localities the skilled mason is becoming rare because there are not many new apprentices. The cost of construction field labor is almost prohibitive for any kind of hand work. The desire for large glass areas lends itself to prefabricated panels. Such prefabricated walls are economical and they are an expression of the modern idiom in construction.

THE EXTERIOR WALL

The curtain wall to be described here is purely a building skin. Unlike the concrete, or steel, or masonry wall it does no structural work and cannot even support itself except on a floor-to-floor basis. The sole purpose of a curtain wall is to enclose the building in a soundproof, weather-resistant envelope. Depending on the ingenuity of the architect engineer, the curtain wall can be built of many materials and in any pattern, subject, of course, to economic considerations.

1E1. *Some Examples of Designed Curtain Walls*

This section will first describe a medium priced, well designed flat wall of aluminum and structural glass with fixed windows (appropriate for a good quality office building). This building is the headquarters of an influential firm. Its exterior was designed to give an impression of dignity, solidity and good taste, without extravagance or ostentation. Its color is natural aluminum and black structural glass. It is continuous, with no projections for columns that are enclosed within the skin. In this example the module of the building is 3'6" so that the vertical mullions are 3'6" on center. These mullions consist of extruded[1] aluminum sections shaped like steel I-beams. They were delivered to the construction site in two-story-high lengths and were fastened to the structural steel at every floor by straps welded to the steel and bolted to the aluminum mullion. Between the mullions at each floor a structural quality glass and aluminum spandrel was fastened and these spandrels run from the head of one window to the sill of the window above. The specifications carefully describe the alloy to be used for the aluminum members by ASTM standards. The spandrel panels are backed with a sound-deadening resilient hydrocarbon resin base material that is spread on to a thickness of about ⅛ inch, and is sag-resistant to a temperature of 300°, and does not become brittle at a temperature of −30°. The panel is further backed with glass fiber semirigid board to prevent heat transmission.

Usually, in all curtain wall designs with fixed glass, the subcontractor who manufactures and erects the wall assumes responsibility for the glazing. Although the glazing is done by another subcontractor it is such an important part of the wall that responsibility for the integrity of the entire wall must be taken by one contractor who will guarantee that the entire wall will fit and work together.

In this example, the glass, which is 3'6" × 6'4", is of "A" quality ³⁄₁₆-inch thick, heavy grade sheet glass and conforms to Federal Specification for Glass. The ³⁄₁₆-inch glass can be used in this case because of the small size

---

[1] An extruded shape is formed by reducing the material (in this case aluminum) to a semiplastic state and then forcing it through a die to form the desired shape.

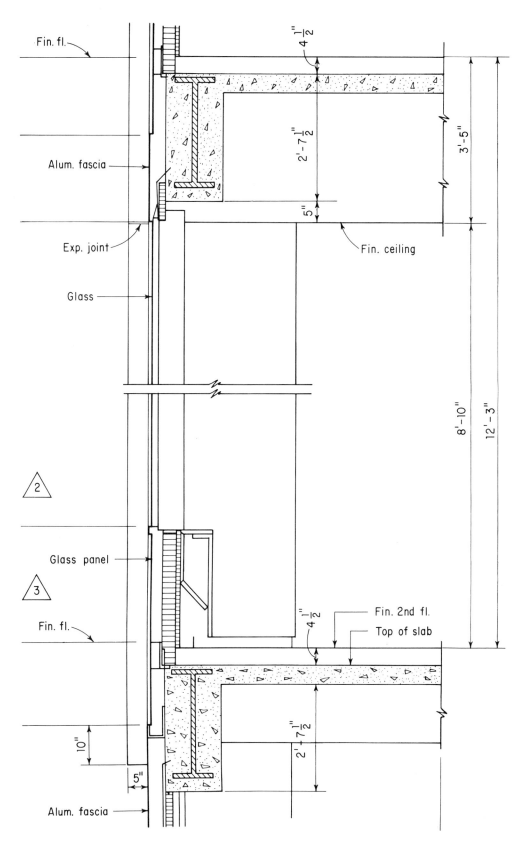

Section of a Curtain Wall.

203

THE EXTERIOR WALL

of the window. For sizes that may be more than 12 inches or so larger in either direction, it is recommended that ¼-inch polished plate glass be used.

Because a curtain wall is built of several materials each with a different coefficient of expansion, and is assembled by many component parts, great care must be exercised to keep it air-tight and water-tight. The sectional drawing shows how carefully the members are fitted together and the specifications carefully spell out the requirements for sealing and caulking all joints and especially those between different materials. First, the specifications call for a primer to seal off leeching of alkaline material from any masonry or concrete surfaces to which a sealant compound may be applied. Such leeching will badly discolor any skin material. The sealant itself is a nonstaining synthetic rubber conforming to a Federal standard and is of the type that does not become brittle. Continuing elasticity in the sealant insures the weather-tightness between two different kinds of materials. The specification also calls for a mastic compound to caulk the joints around the heat insulating board and the metal frame into which this board is fitted. This caulking prevents the flow of air and acts as a vapor seal to cut off the continual condensation that could occur were air allowed to flow freely between the insulating board and the exterior surface material.

A second example of a metal and glass curtain wall illustrates the architect engineer's and owner's desire to produce a striking building in a city in which heretofore there were not many examples of striking architecture. The aluminum skin for this 34-story building is designed in deep and bold patterns around projecting columns and projecting sills. The windows are set back from the face of the columns and sills so that these projections not only provide shade from the sun but also give an effect of deep shadows that change in shape and quality as the sun's direction changes. Because most of the buildings in this city were of light colored stone or concrete, the skin in this case was specified to be of dark bronze aluminum and bronze colored glass. The large aluminum sheets are specified to be of heavy enough gauge to prevent "oil canning" or buckling under wind pressure. Their color is specified as a hard coat anodized finished controlled alloy of dark bronze. The anodizing process on aluminum is an electrolytic process by which a color is bonded to the aluminum. Such anodic coating is subject to ASTM standards and all major manufacturers observe them. Because it is important in a large structure to avoid any wide variation in color, especially in this dark color, the specification calls for very careful color comparison between aluminum sheets and allows only a small color differential. The color testing is done in accordance with the National Bureau of Standards formula and uses a Photo-Electric instrument for comparison. As in the first building mentioned, the sound deadener, the sealant, and the mastic caulking compound are carefully specified. Most specifications also call for a plastic protective coating on the aluminum to shield it against rubbing and scratching until it is erected and the building is completed.

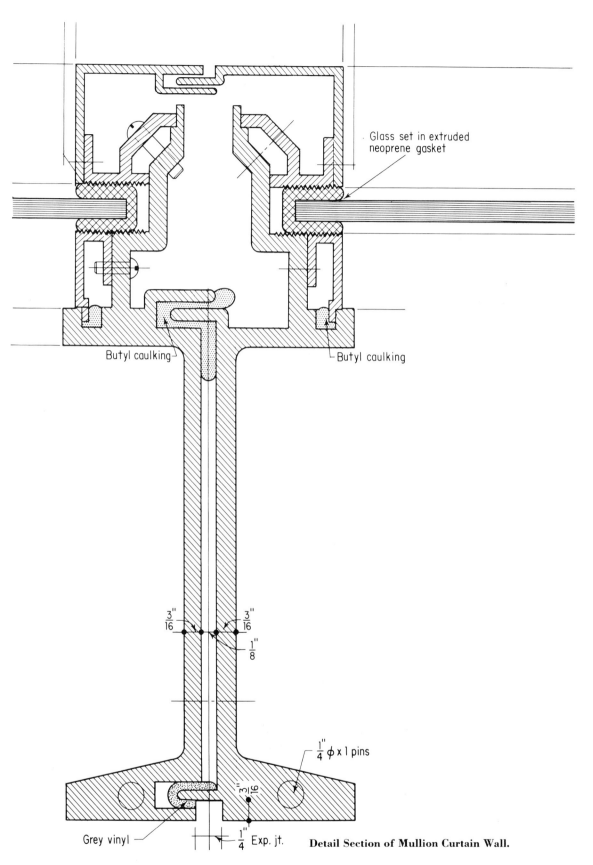

Detail Section of Mullion Curtain Wall.

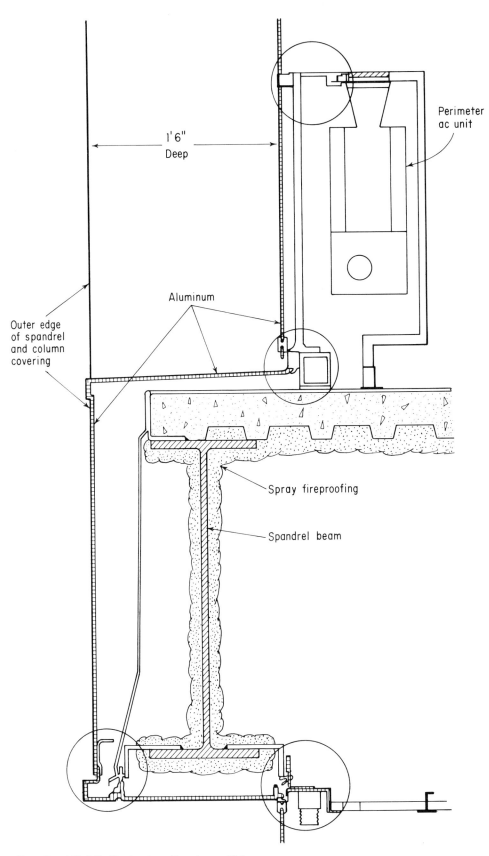

Details of Boldly Projecting Aluminum Skin.

THE EXTERIOR WALL

1E1a. *A typical specification for aluminum and glass*

Before going on to the next examples of designed-to-order building skins it might be well to quote a typical specification for the erection of an aluminum and glass curtain wall. The first requirement is that insofar as is practicable the fabrication, assembly, and fitting of the work be executed in the shop and that work that cannot be shop-assembled be given a trial fit before being sent to the job. The joints, miters, and corners must be accurately machined and rigidly fitted together with only hairline cracks visible. Screws are either stainless steel or bronze. The flatness of the aluminum sheets is held to within $\frac{1}{32}$-inch deviation. The careful architect engineer watches the delivery and erection of the curtain wall very closely.

1E1b. *Description of masonry and glass wall and specification*

The curtain wall to be described here was designed for a 34-story building in a city with a mild climate. The owner and architect engineer desired a building that was impressive in its expression but which at the same time would give a feeling of lightness and warmth. The building wall consists of a series of vertical mullions of precast architectural concrete. Although these mullions are quite large (as shown in sketch) and are spaced on 4′8″ centers, the building is so large and the color of the concrete so warm that the effect is one of airiness. Between the mullions the plans call for bronze colored spandrel glass and bronze anodized aluminum frames to hold bronze colored glass.

The specifications for the precast architectural concrete is worthy of interest and should be carefully noted because the erection of this particular skin is completed and it has turned out extremely well. The color is even and warm and the erection of these heavy mullions was performed without any trouble in the connections or the fitting. The specification stated that no manufacturer with less than five years of experience in the manufacture of such exposed aggregate precast architectural concrete could bid the work. Upon the award of the contract and before any work could start the manufacturer had to submit large samples for approval. The materials to be used for the "face" mix were carefully specified by size and color. The cement was of the stainless[2] variety. The fine aggregate was Silica white sand all of which could pass through a number 8 sieve (fairly coarse). The coarse aggregate was crushed buff Texas limestone ranging in the size from $\frac{3}{8}$ inch to $1\frac{1}{4}$ inch with most of the stone about $\frac{5}{8}$ inch. The backup material consisted of regular Portland cement, sand, and $\frac{1}{4}$- to 1-inch coarse aggregate.

---

[2]Stainless cement, as its name implies, does not stain marble, or granite, or any light colored stones with which it comes into contact. Most cements do. Stainless cement is more expensive.

THE EXTERIOR WALL

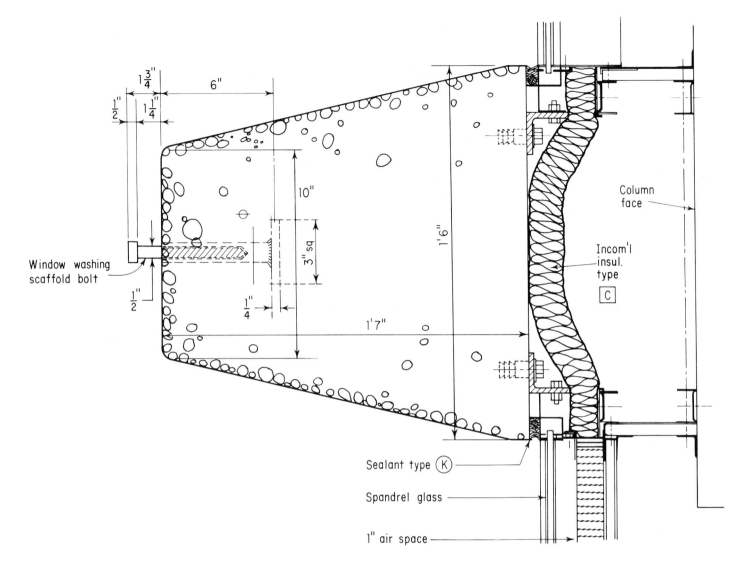

Section of Precast Concrete Mullion.

The batching plant where the concrete was mixed was inspected to make certain that there were facilities for accurately weighing the ingredients and to make sure that the ingredients could be thoroughly mixed before the carefully measured water was added. The specification called for reinforcing by both steel mesh and steel rod and such reinforcement had to be carefully buried in the cast member far enough from the exposed surface to prevent any possibility of the seepage of water and consequently of rust. The concrete mixture was placed into heavy metal forms in a fairly dry state (only 2-inch slump) and was then thoroughly vibrated into place by bouncing the form up and down on a heavy steel covered concrete floor. (The author still remembers the deafening noise and that people living almost a mile away complained. Sound baffle walls had to be built.)

When the concrete had attained its initial set it was kept wet by a fog spray mist for several days and then air-dried for 21 days. The tolerances were kept to a maximum of ± ⅛ inch in 8 feet. Before being shipped to the job each surface was cleaned with a dilute solution of muriatic acid and then washed thoroughly and given two coats of sealer to prevent any penetration of moisture which could reach the steel reinforcement. Each mullion and each panel were provided with carefully placed bolts which were positively attached to the heavy reinforcement. The end result of all this care was that almost 2000 mullions each weighing 8600 pounds were erected in 99 days with no fitting problems and there were almost no variation in color or texture. The erection specification, which was strictly adhered to, called for bolting the units to angles welded to the structure by certified welders and for the cleaning and coating of all steel connections with a cold galvanizing compound.

Careful procedure in manufacturing and erecting architectural concrete resulted in a job of which the architect engineer and owner could be proud. There is no reason why such a procedure should not be adhered to for any architectural precast concrete or cast stone to be used for a building wall. Too often, because of penny-wise economy such material is rushed to a job without proper reinforcing or curing and too often the material starts to crack, sometimes within a few months of erection. As a result of weather erosion, such cracking can become a serious maintenance problem.

Another matter that should be mentioned here but which of course relates to any part of a construction job, is scheduling. Because the mullions mentioned above were so heavy and so large they had to be hoisted by a large crane. The lower floors were serviced by a truck crane and the upper floors by a roof crane. Because the material could be delivered from only one face of the building, the erector had to install a monorail on the exterior of the building, one half-way up, and one at the top to receive the mullions and panes from the cranes and to distribute them around the perimeter. As the material was being delivered there was always one truckload arriving as the preceding one was being unloaded. One team of two men attended to the hoisting while another team rigged hoisting cables on the load that had just arrived. The erection was performed in record time and was, of course, a financial success.

The curtain wall described above consisted of the precast concrete mullions, and of structural glass spandrel panels, and windows between them. The specifications for the material and setting of windows will be described in a later section. The specification for structural glass, which is widely used in exteriors is as follows: Structural or spandrel glass shall be heat-strengthened ¼-inch polished plate glass with ceramic material fire fused on its back surface to become an integral part of the glass and to provide an opaque appearance. This glass is manufactured by several companies and can be procured in designed sizes and in a wide variety of colors. In this curtain wall the glass spandrel is fastened within an extruded color-anodized

THE EXTERIOR WALL

aluminum frame which also holds the window glass above and below the spandrel and in turn is fastened to the structural frame of the building. All joints between the various materials that make up the wall are carefully sealed with special compounds that preserve their elasticity and adhesiveness under all conditions.

1E2. *Glazing Practice*

The glazing of a curtain wall is performed by a contractor who should be responsible to the general curtain wall contractor for the proper setting and weather-tightness of the installation. Most curtain walls use fixed glass (windows that cannot be opened) in their construction and because this installation has to work in harmony with all the various materials of the wall, the specifications for the material and the method of glass installation are carefully drawn. The architect engineer can choose from several colors of glass to fit the design. The thickness of the glass is determined by its size and has been set by Federal specification and tables of strength for various wind loadings. While local codes do not specify glass strength, they do follow the national and trade codes.

1E2a. *A typical specification*

A specification for glazing would read as follows: Glass shall be of domestic manufacture of size and thickness indicated on drawings and each light must be labeled with the manufacturer's name and its grade. Clear glass shall be polished plate that has been twin ground. The colored glass shall also be polished twin-ground plate glass. Field cutting, especially of colored glass which is heat treated, is either forbidden or must be done under careful supervision to avoid temperature stress breakage during cutting, or the setting up of strains that may result in damage after installation.

The setting of the glass into the frame has to be done very carefully. The joint must be tight enough to exclude noise and weather but must have enough play to avoid the rubbing of the glass against metal when the building sways or vibrates and settles as all buildings do. In all cases of fixed windows the specification calls for a neoprene[3] gasket manufactured to the exact size of the frame and into which the glass is set. The gasket is usually specified to have carefully molded corners for exact fit. The specification also calls for the glass to be held in place and kept apart from the frame by small blocks of hardened neoprene which reinforce the glazing gasket. This last item is very important. The writer knows of a building in which several lights of glass at various locations started to crack (mostly diagonally). It was determined that the glass was cracking because it was rubbing against the metal frame. The frame and glass were tested for squareness and were

[3]Neoprene is a synthetic rubber of superior weather resistance which can be manufactured in various hardness. It retains its elasticity for long periods.

found satisfactory. It was finally determined that in certain sections of the building, where the wind loading came from an angle and caused side vibration, the glass was slipping diagonally in the frame and rubbing against it. The original specification called for neoprene setting blocks on the bottom of the glass and this is normally sufficient, but in this instance blocks will have to be placed on either side of the glass as well, to keep it from slipping sideways in the frame.

1E3. *Prefabricated Curtain Walls (Pivot Windows)*

In smaller structures, where because of the size and the small quantities involved it becomes uneconomical to install a specially designed exterior wall, the architect engineer has a wide range of choices in prefabricated curtain walls that are factory-assembled and come to the job ready for field installation. The industrial exterior wall, which can be obtained in many materials, colors, and sizes has been previously discussed in this chapter. This wall is essentially for large unbroken expanses. For the small office building, medical building, or small apartment house, such walls come pre-assembled in combinations of colors and materials. The vertical mullions that hold the wall to the structure are usually of aluminum but the spandrel panels can be of steel or aluminum with a baked-on acrylic enamel, or of structural glass, or reinforced fiber glass. All such spandrel panels are of sandwich construction with laminations of insulating material, sound deadening material, and vapor barriers and can be finished on the inside with a hard board that can be painted. For the one-story building there are some curtain walls that are load bearing. The vertical mullions can be large enough to hold up a roof structure as well as serving to frame the wall structure itself. Each combination of color, and material, and the interior module which it expresses should have a relationship to the size and shape of the structure and if possible to its use.

2. MOCKUPS, PERFORMANCE STANDARDS, AND TESTS

The architect engineer, when designing a curtain wall, always specifies a mockup, which is built of the actual materials to be used and shows in actual full-size proportion how the various materials will look and how they will fit together. Such mockups can vary in size from two or three vertical and horizontal panels to an entire small building. These mockups have another purpose, since they can be tested for weather resistance in the equivalent of a wind tunnel.

Before being tested these mockups are built to the specified performance standard. Such a typical standard would have provisions for thermal movement providing that the wall be constructed so as to allow for the movement of the component materials within a temperature range of 150°F without causing the opening of joints or undue stress on the fasteners or the

THE EXTERIOR WALL

glass. The structural strength of the entire wall is specified to withstand a wind of 110 miles per hour[4] with the following allowable deflections: For a metal framing member, not more than $\frac{1}{175}$ of the clear span of the member or ¾ inch, whichever is less, in a direction at a right angle to the wall and not more than 75% of the design clearance in a direction parallel to the wall.

The neoprene gasketing has to remain pliable to $-40°F$ and show its weather resistance by no visible deterioration after being exposed to ozone at 100°F for 100 hours and at 20% elongation. It must not rupture or even remain permanently elongated after being stretched to 175% of its length. The other materials such as anodized or enameled aluminum or the sealants have to withstand exposure tests and aging tests as prescribed by ASTM standards.

When a piece of full-size sample wall has been built and approved for color and design, and all the component parts have been tested and certified to meet performance standards, then the entire assembly is tested as an entity. There is a standard testing procedure set up by the National Association of Architectural Metal Manufacturers (NAAMM) and there are *certified testing agencies that perform the work*. The test is performed by a propeller type wind generating device with a propeller diameter of at least 10 feet. The propeller must be capable of producing a wind velocity of over 110 miles per hour at the face of the test section. A sprinkler is mounted in front of and over the propeller. The sprinkler must be capable of delivering the equivalent of 4 inches of rainfall per hour per square foot. The test wall must be able to withstand this very heavy rainfall at the high wind velocity produced. The water infiltration and the wall deflection are measured. This wind and water testing is very important. It shows flaws in a design or a material *before* the actual curtain wall is built. The architect engineer must insist on such actual tests for a specially designed wall and for a certification that such tests have been satisfactorily passed if he is using a stock prefabricated wall.

3. SOME EXAMPLES OF SPECIALLY DESIGNED WALLS

When the architect engineer is given the latitude to design a particularly unique exterior he can produce some really striking and beautiful effects. Even in a commercial building, he can design an exterior of special steel, or architectural concrete, or mirror glass that is pleasing in color and in proportion, monumental in effect, and economical to build. When he goes further he can produce a building such as the Beinecke Rare Book Library at Yale University, or Philharmonic Hall in Lincoln Center in New York City,

---

[4]Wind loading tests refer, of course, to the wind velocities characteristic of the local community.

THE EXTERIOR WALL

or the Ford Foundation building in New York City or the Virginia National Bank Building in Norfolk. These buildings are expressions in steel, and stone, and glass, and metal. The Stiles and Morse Colleges at Yale University in New Haven are a unique expression in masonry. (See the photograph on page 196.)

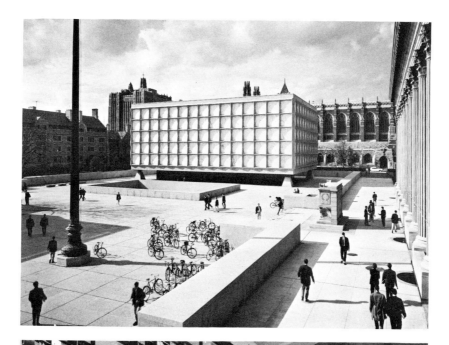

Beinecke Rare Book and Manuscript Library. Courtesy Ezra Stoller Associates, Inc. Skidmore, Owings & Merrill, architects.

THE EXTERIOR WALL

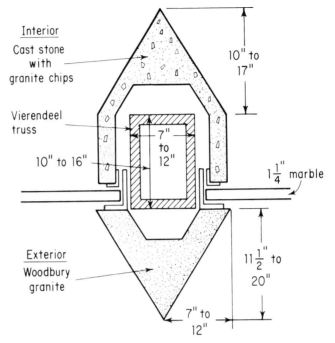

Section Typical Vertical And Horizontal Mullion

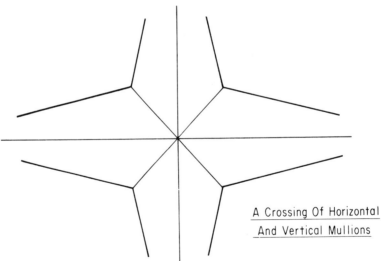

A Crossing Of Horizontal And Vertical Mullions

Beinecke Library.

The Beinecke Rare Book Library consists of a building within a building. The inside structure is a structural entity built of steel, and glass, and totally encloses the rare books and manuscripts which are stored in stacks under a constant temperature of 70°F and 50% relative humidity. The outer structure consists of a series of Vierendeel trusses, each section being 8'8" by 8'8" with all welded connections. The diagram shows a section of a truss and illustrates how the cast stone and granite facings are hung on the trusses and the marble set in the truss. The marble is 1¼-inch-thick Vermont marble which has tones of white, gold, and yellow. It is translucent and glows on the interior during the day and on the exterior after dark. The exterior skin serves to cut out the damaging rays of the sun and makes the task of air conditioning the interior stacks much easier.

Philharmonic Hall is an example of a practically free-standing curtain wall. The front entrance lobby rises to the full height of the structure. The wall of the concert hall is independent of the front wall and contains promenades at the front and sides to enable the audience to walk completely around three sides of the hall. The side promenades extend from the concert hall shell to the side walls. The free-standing front columns are clad in travertine marble. The lobby and front promenades, especially after dark, present a glowing scene of movement and color.

Philharmonic Hall, Lincoln Center, New York. Photograph from Harrison & Abramovitz, Architects. Photograph above Courtesy Ezra Stoller Associates, Inc.; Photograph on Right Courtesy Bob Serating.

THE EXTERIOR WALL

The Ford Foundation building presents a front wall of structural steel and glass which is free-standing and serves as a frame for the interior terraced gardens and the office structure that lies behind the gardens. The large sheets of plate glass which constitute almost the entire wall are framed into the very light looking structural steel frame and the entire front wall looks airy and graceful—almost as though it weren't there.

Ford Foundation, New York City. Kevin Roche, John Dinkeloo and Associates, Architects. Courtesy Ezra Stoller Associates, Inc. for the Ford Foundation.

THE EXTERIOR WALL

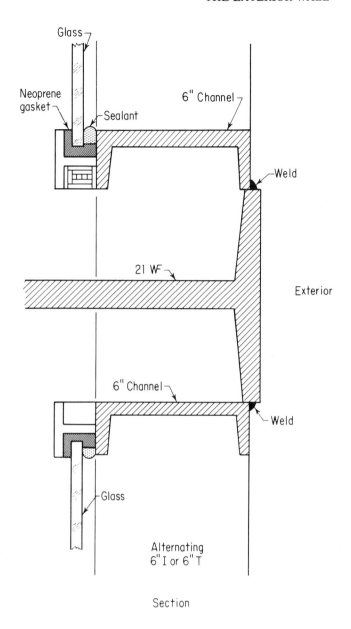

Ford Foundation, N.Y.C.   Exterior Wall Construction

A precast structural concrete exterior wall presents bold, strong lines and gives an expression of the structure of the building in the Virginia National Bank Building. The structural exterior walls consist of a series of cast concrete of tees each 12′2″ high and 14′0″ wide. The architect engineer designed all the tees to be the same size although the bottom ones carry the entire structure and the top ones only carry their own floor loading and the roof. The difference is in the reinforcement. The bottom tees are almost solid reinforcing rods and the top ones are very lightly reinforced. The single

217

THE EXTERIOR WALL

size was used to emphasize the straight vertical lines of the columns, and also because forms are expensive, and in this case only one size was required. The tees are placed one above the other and are fitted together by male and female steel sockets which are welded to the reinforcing rods. As each tier of tees was placed the structural concrete floor was poured, thus forming a homogeneous working structure.

Virginia National Bank, Norfolk, Virginia. Skidmore, Owings & Merrill, Architects. Courtesy Ezra Stoller Associates, Inc.

THE EXTERIOR WALL

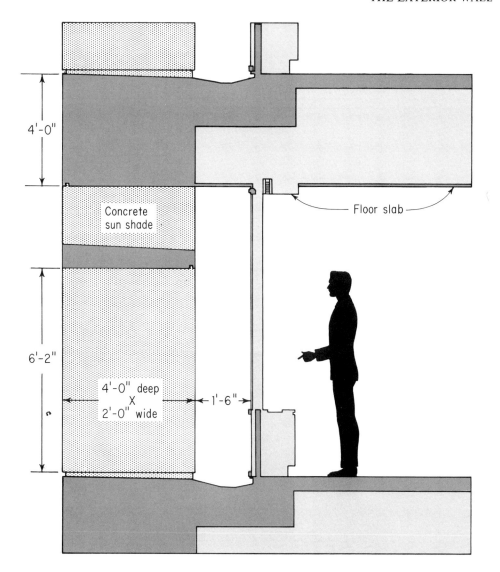

Exterior Wall Section

Virginia National Bank.

There are many other fine examples of imaginative and beautiful exterior walls. Two that are shown here are the Phoenix Mutual Life Building in Hartford, Conn., which presents an all glass wall in an uncommon shape and the Chapel at Brandeis University, which is an outstanding example of the use of masonry and glass in a simple dignified form combined with the use of water as a simple reflector of good architecture.

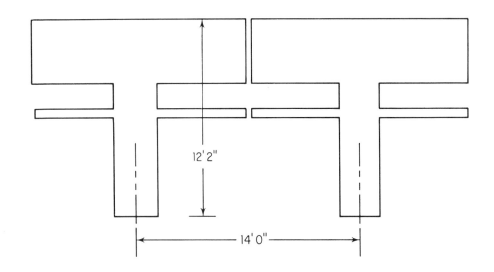

Precast Concrete Tees Used As Bearing Wall

Virginia National Bank.

Phoenix Mutual Life Building, Hartford, Connecticut. Harrison & Abramovitz, architects. Joseph W. Molitor, Photography.

THE EXTERIOR WALL

Jewish Chapel at Brandeis University. Harrison & Abramovitz, Architects. Courtesy Ezra Stoller Associates, Inc.

CONTENTS

1. THE IMPORTANCE OF THE FINISHED FLOOR
   A. The Private Residence
   B. The Small Office or Apartment Building
   C. Special Floors
   D. Heavy-Use Floors
   E. Large Office Buildings
      1. Resilient flooring
      2. Terrazzo floors and quarry tile
      3. Raised floors
      4. Carpeting
2. CEILINGS—A GENERAL DISCUSSION
   A. The Hung Ceiling
      1. Composite ceilings
      2. Luminous ceilings
      3. Special-use ceilings
      4. Acoustic plastered ceilings

13

FLOORS AND CEILINGS

1. THE IMPORTANCE OF THE FINISHED FLOOR

The finished floor in any structure requires the close attention of the architect engineer. It plays a very definite part in the decorative and architectural character of a structure. It should also be practical. It should be durable, simple to keep clean, and able to withstand whatever use it is put to. There are dozens of floor treatments and surfaces available. The architect engineer should consult the owner as to a choice of floorings and must determine which one will do the best job within the economics of the situation whether it be for its wearing or aesthetic qualities.

1A. The Private Residence

The choice of the finished floor in a private residence can range from asphalt tile on a concrete slab to wall-to-wall carpeting. Middle- to upper-grade private residences usually have a rough underfloor of 8-inch wide tongue and groove pine boards, preferably laid diagonally for structural bracing, or parrallel to outer walls which is not as satisfactory but is being used more and more because of labor costs. Over this is laid a layer of felt or tar paper and the floor is finished with hard wood, usually tongue and groove red oak, laid in strips. (The finished floor is then sanded and given several coats of filler and wax.) This finished floor can also be maple, wal-

FLOORS AND CEILINGS

nut, or beech depending on the owner's taste and pocket book. The kitchen and bath are usually floored first with at least ¾-inch-thick plywood which is machine sanded and over which a resilient floor of vinyl, linoleum, or asphalt tile is then laid.

Within these conventional finishes the choice of flooring varies widely. In a development house built on a concrete slab it is usual procedure to place a sheet of polyethylene (very thin plastic not unlike cellophane) under the slab to prevent seepage of ground moisture and then to cover the entire slab in every room with resilient flooring of asphalt tile laid in troweled-on mastic (a bituminous based, waterproof adhesive). When the houses are of better quality the resilient floor may be of vinyl asbestos which is more expensive than asphalt tile, and has a wider range of color, and wears better. In a high-quality residence designed and built for a specific owner and constructed on a flat slab, the architect engineer can suggest wood parquet flooring in the living room and bedrooms, and vinyl tile in the kitchen, and vinyl or ceramic tile in the baths. Wood parquet consists of squares made up of a number of thin parallel strips of wood bound together to make a square 9″ by 9″ by 5/16″ thick. These squares can vary in size. They are laid in mastic in alternate patterns to give a checkerboard effect. The wood may be red oak, maple, walnut, or other hardwood. In the kitchen a ⅛-inch-thick vinyl tile, sheet vinyl, or sheet linoleum offers many excellent color and pattern choices. Vinyl gives a clear-colored, resilient, hard-wearing floor which requires very little cleaning and is resistive to most kitchen spills. It comes in many handsome patterns and can imitate terrazzo, marble, brick, or tile. Consideration should be given to ceramic tile flooring in bathrooms. Many

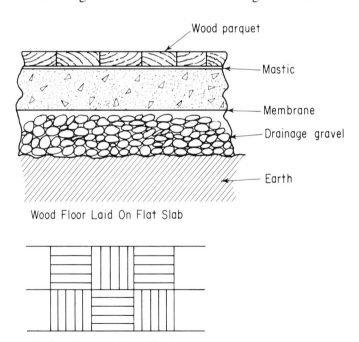

Wood Floor Laid On Flat Slab

Typical Parquet Floor Pattern

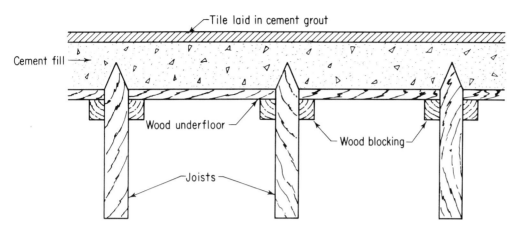

Typical Base for Ceramic Tile Floor.

color combinations and patterns of mosaic floor tile or matt-finished larger tile are available. They are usually laid in a thin cement grout and last for almost a lifetime. Mosaic ceramic tile comes in sheets of 1' × 2' or 2' × 2' and is laid face down into the grout after which the backing is removed. In the private residence where joists hold the floor the wood parquet flooring and vinyl flooring can be set in mastic over a heavy plywood base that is at least ¾ inch thick. The ceramic tile is laid over a cement base which is laid over a depressed floor as shown in the diagram, or it can be laid in an epoxy glue over heavy plywood (which is not as good).

Wall-to-wall carpeting can be laid over ¾-inch plywood which should first be sanded. If the floor under the carpeting is not finished, the owner can never take the carpet with him, and he should consider this before making his decision. In all cases the concrete slab and plaster walls must be thoroughly dry before any finished flooring is even brought into the house.

1B. The Small Office or Apartment Building

The flooring in a small office building is usually of the resilient type such as vinyl asbestos or asphalt. If the floor slab is light weight concrete or any of the many types of precast materials it must be laid on dense concrete at least 1 inch thick which must be laid over such slab. If the floor slab is of poured conventional concrete it must be carefully troweled to a smooth finish and must be thoroughly dry before the floor tile can be laid. In the small apartment building the floor can be of vinyl asbestos laid over concrete or a double-laminated plywood floor system, but in the better buildings it is either red oak, tongue and groove strip flooring laid over wood, or metal "sleepers," or parquet flooring laid in mastic directly on the concrete floor slab or the plywood underfloor. The flooring can be impregnated with plastic for long-wearing qualities.

FLOORS AND CEILINGS

The sleeper is a strip of wood laid over a concrete slab to which a flooring board may be nailed. The sleeper may also be a steel channel to which the flooring board may be fastened at intervals. Either separates the finished flooring from the concrete slab and gives excellent soundproofing quality. If the floor system is made of wood such as double-laminated plywood, then the flooring is nailed directly to it.

1C. Special Floors

There are special-use floors to suit almost every activity. For gymnasiums there is a shock-absorbent floor which consists of two layers of plywood laid over a shock-absorbent padding. This padding is laid on a concrete slab that has been waterproofed with a top coat of membrane waterproofing. The top layer of plywood is covered with a hard maple floor. The maple is carefully sanded and finished with several coats of a penetrating filler and possibly a coat of nonslip wax. Instead of double plywood the underfloor can consist of a single layer of plywood laid over the resilient pads and then wood sleepers on which the maple is laid. Such floors can also be used in assembly rooms and auditoriums. Hospital floors are a problem because they must be resistant to chemicals, easy to keep clean, able to withstand heavy traffic and must be quiet. Resilient rubber flooring is often used. It is resistant to stains and scorch. It does not shrink and therefore there are no cracks to hold dirt or germs. Nor does it cause static build-up, which makes it practical for use in an operating room. It can also be flooded with a germicidal solution.

There are some floors that can be troweled on over a concrete slab. Some of these are chemically inert and therefore resistant to chemical spillage. There are floors that are conductive so that there will be no static build-up which could cause an explosion. For schools, churches, and libraries there is natural cork flooring that is laid in mastic over wood or concrete underfloors. This is impregnated with color and protected by a vinyl coating. It comes in ⅛-inch or ¼-inch thickness. It is very quiet, looks beautiful, and is expensive. There are combinations of cork and plastic that can be troweled on over wood or concrete or there are combinations of colored plastic chips in a clear resin which make good-looking, long-wearing floors for corridors in small office buildings or apartment houses.

1D. Heavy-Use Floors

Many floors are available for the factory or warehouse where heavy loads are transported and where there may be spillage of oil, grease, chemicals, or heavy abrasion by metal chips or falling objects. There is a floor available that can be built up of a troweled-on resin containing chips of hard material and laid in several coats over a concrete or heavy wood floor. Another hard-wearing floor is built of a concrete mixture containing a large proportion of coarse hard aggregate which can be vibrated into a very dense ¾-inch-thick mixture over a concrete floor slab and then steel-troweled to a smooth hard-wearing surface. An epoxy coating can be troweled over a concrete or wood floor to make it resistant to chemical spillage. There is also a wood block floor that consists of pine blocks laid on edge. These blocks vary in size from 2½" × 5½" to 4" × 8" and are either 2 or 4 inches thick. They are laid in hot tar over a concrete base and then covered with hot tar. They make an extremely sturdy floor that is resistant to almost anything. There are many more floor treatments and each has properties to fit the job it has to do.

1E. Large Office Buildings

1E1. *Resilient Flooring*

The flooring in a major office building is generally of resilient tile laid in troweled-on mastic adhesive on a finished cement floor. Because so much of it is used, a typical specification for such a floor follows: Resilient tile shall be of approved manufacture ⅛ inch thick and 9 inches (or 12 inches) square. It shall be factory waxed and of uniform color. Vinyl asbestos tile shall conform to Federal Specification L-T-751 and shall consist of a thoroughly blended composition of vinyl plastic resins, asbestos fibers, alkali resisting pigments, and fillers. Vinyl tile shall conform to and shall consist of vinyl resins, alkali resisting pigments, and other plastic compounds. Rubber base shall be 3 inches high and not less than ⅛ inch thick. Subfloors shall be level, clean, and thoroughly dry. All rooms to receive resilient floors shall be maintained at a minimum temperature of 70°F at least 48 hours before and 48 hours after the floor has been laid. Adhesives shall be applied evenly with a notched trowel. The tiles shall be installed with tightly butted joints and edge strips to be supplied where the tile abuts cement-finished, terrazzo, or other floors. The tile must then be cleaned and protected.

Because of the cost involved, most resilient floors in good office buildings are of vinyl asbestos rather than pure vinyl. Some office buildings use asphalt tile which costs less than vinyl asbestos and has neither the color nor the wearing qualities. Vinyl flooring may be used in bank or other specialty spaces.

1E2. *Terrazzo Floors and Quarry Tile*

Another flooring material used extensively in office buildings and public spaces is terrazzo. This is a mixture of stainless cement, fine sand, and marble chips. It presents a very hard smooth surface that is easily cleaned

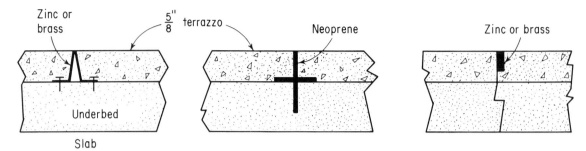

Examples of Expansion Joints in Terrazzo.

and is very pleasing and luxurious in appearance. Terrazzo is now being used as exterior paving in large plaza areas and on sidewalks. It has always been used in building lobbies, and in public corridors, and elevator landings. It can withstand a tremendous amount of foot traffic with almost no signs of wear. Terrazzo is specified to be a mixture of marble chips of selected colors and texture that is set in a matrix of cement and sand. Very often the specification calls for the coloring of the matrix to complement the color of the marble. Terrazzo is always set over a concrete slab. An underbed of mesh-reinforced, cement mortar at least 1⅛ inch thick and consisting of Portland cement and sand is laid over the concrete slab. After this coat has been tamped to a firm and level surface a series of brass or zinc divider strips is pushed into it to form a pattern for the finished terrazzo and to act as expansion joints. As the diagram shows there are many varieties of expansion or divider strips. Almost all contain a neoprene core and are made of corrosion-resistant metal. After the expansion strips are placed in the desired pattern the terrazzo itself is mixed with just enough water to make it plastic. It is then "screeded" over the divider strips which at this point project ⅝ inch from the underbed. (*Screeding* is the rough leveling of a plastic surface by running a level edge over it in a backward and forward motion as the edge advances.) The screeding forces the terrazzo to fill in all the voids between the divider strips. The entire surface is then thoroughly tamped and rolled to present a level surface and a homogeneous mass. The terrazzo is then cured for at least seven days by covering it with a layer of wet sand which is kept wet for the entire period. After the curing the terrazzo is ground by a grinding wheel with a heavy grit embedment. It is then reground using successively finer stones and a cement grout of the same color as the matrix. A great deal of water is used during all these grindings. It can be brought to as high a polish as desired. Such high polish is not advised for outdoor terrazzo which can get quite slippery. Terrazzo can also be acid etched to roughen it and to provide a particular texture, or it can be sprinkled with an abrasive material to form a positive nonslip surface. When terrazzo is installed over a corrugated structural steel floor the minimum thickness of the terrazzo and the underbed on which it is placed must be at least 2 inches instead of the 1¾-inch thickness required over a solid concrete slab. The architect engineer

FLOORS AND CEILINGS

must remember to clearly mark his drawings where terrazzo or another floor that is thicker than usual is to be installed because the structural floor must be finished at a lower level to allow for it.

Terrazzo also comes in a plastic mixture. This is called *epoxy, polyester,* or *resin.* It is troweled on over a heavy wood or concrete base and then lightly sanded to bring projecting marble chips to a smooth surface. It is useful in many areas such as corridors of office buildings or in private homes in baths and kitchens.

Precast polished terrazzo is used extensively for stairways. In an ornamental stairway it can be used as a heavy tread with or without the riser, or in a single piece as a combination tread and riser. It can be obtained with abrasive nonslip surfaces. It also comes in a tile form.

Quarry tile, which is a hard burned flat clay tile usually terra-cotta colored is often used as a finished flooring material. These tiles are laid in cement grout and can be used inside or out. Many patios, plazas, and walk-on roof terraces are paved with quarry tile. It is weather and abrasion resistant and pleasing in appearance.

1E3. *Raised Floors*

The advent of computer centers, where the various parts of the computer system must be in certain relationships with each other and where the entire set-up may be changed at any time, has brought into use the raised floor or pedestal floor. There are many variations of this but they all perform the same function. They consist of a complex of steel pedestals that hold metal pans or tiles away from the structural floor so that the space between may be used for the mass of electrical cables required to connect the component machines to the source of power and to each other. The aluminum pans are usually covered with vinyl. They can be lifted for access whenever there is trouble or new connections have to be made.

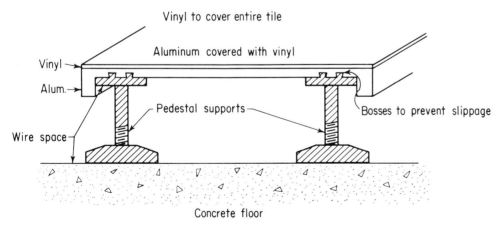

Typical Arrangement for Pedestal (raised) Floor.

FLOORS AND CEILINGS

1E4. *Carpeting*

Carpeting of publicly used floors in office buildings has recently become quite popular and the architect engineer and owner should be familiar with its qualities and use. It gives the office building tenant and his visitor a sense of quiet luxury which helps greatly in competitive rental markets. Carpeting is used mostly in upperfloor elevator landings or lobbies and in upperfloor corridors. This section will discuss three of the many kinds of carpeting. *Wool* is stain and soil resistant or can be cleaned from most stains that might occur in a corridor. A 44-ounce-weight (fiber content) wool rug will last from 7 to 10 years and will look good during its entire life. The natural wool fiber takes dyes very well and gives a lasting soft luxurious look. *Antron nylon*, which weighs 28 ounces per square yard, has a better resistance to stain than wool, and because of its opaque fibers is almost as soil resistant as wool. It wears well for at least 10 years. Its colors are bright and shiny and it has a synthetic and flat look. It is slightly cheaper than wool and generally is excellent carpeting. *Nylon* weighing 20 ounces per square yard is stain-resistant but soils badly. It can be easily cleaned but this must be done much more often than for other fibers. It wears indefinitely and is cheaper than the others, but its colors are comparatively dull and it looks flat. All carpeting generates static when walked on in dry cold weather. It should be sprayed in cold weather with an antistatic spray. All carpeting is laid over a carpet cushion or undercarpet, usually of 50-ounces-per-square-yard felt or cushioned synthetic rubber. Serious thought must be given to the use of carpeting, especially in competitive situations.

2. CEILINGS—A GENERAL DISCUSSION

A discussion of ceilings will be confined to the hung ceiling that is common to all major office buildings and to many smaller structures. All other ceilings are simply plaster or paint on the underside of the concrete floor or sheet rock, lath and plaster, or other material that may be used to present a smooth flat surface to cover the floor system above.

The hung ceiling is the device invented by an architect engineer whereby a space is created between the underside of a structural floor and the upperside of the ceiling of the floor below. This space, which may vary from a few inches under a heavy girder beam to 3 feet or more in the open spaces between beams, is crammed with air conditioning ducts, air mixing boxes, damper controls, reheat units, electrical conduits, plumbing piping, and the upper part of flush ceiling-mounted lighting fixtures. In addition to this it often acts as a plenum chamber for the return of exhaust air.

2A. THE HUNG CEILING

The hung ceiling is created by the suspension of the finished ceiling of the floor below from the structural floor above. If the floor is concrete, steel straps are bent over reinforcing rods and project from the underside of the concrete. If the floor is corrugated steel, T-fittings are dropped through slots in the steel before concrete is poured over it. Extra fastening into the surface can be obtained by the use of a small explosive charge which shoots a bolt into a concrete surface. The actual suspension is performed by wire tied to the ceiling straps above and to steel channel iron below. The channels hold the entire ceiling and the lighting system. All of the ducts, conduits, and pipes that run through the hung ceiling space are independently hung from the structural slab above. The hung ceiling used most often is fissured mineral tile, cut to a size that meets the modular requirements of the building and set into exposed tee bars as shown on the diagram.

Depending on the module, the specifications for this hanger system call for tee bars of a certain height and width made of cold rolled steel or aluminum. The entire grid of main tees and cross tees is modular in length and width. The cross tees in this case might carry the lighting fixture which would rest on these tees at both its ends. The fixture has to be one module long and can be of any width. A tile rests on either side of it and on the main tee. Often when larger modules, such as 5'0" × 5'0" are used, the lighting fixture is not a full module long or wide and in such cases it rests on concealed steel splines.[1] The ceiling tile itself in this particular case is ¾-inch-thick fiber glass faced with a plastic film. The tile is rated "Class A-

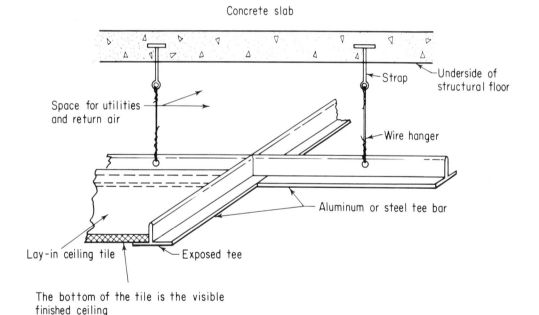

Typical Hung Ceiling. Very Often the Tees are Hung from Steel Channels.

[1] A *spline* is a metal strip that fits into a groove cut into the thickness of a tile and that runs beween tee supports to support the tile.

FLOORS AND CEILINGS

Incombustible" according to Federal Specifications SS-A-118b. This tile is only one of hundreds of patterns which can be chosen. Fissured mineral tile, which is extensively used, is made of mineral fiber such as gypsum and has a sound-absorbent surface in patterns that look like travertine marble, or with random patterns of swirls, or small holes. The usual thicknesses are ⅝ or ¾ inch. In instances where the architect engineer does not want the exposed tees to show he designs a concealed spline hanging arrangement, as shown. This, of course, is more expensive than the exposed tee arrangement because the factory cutting of the tile costs more and the field labor in fitting is more. The tile also has to be handled carefully to avoid breakage of the thin edges. As a compromise there can be an exposed tee running in one direction and concealed splines or tees running in the other. This makes a very pleasing architectural effect. The exposed tee can also be designed to be flush with the tile surface as shown, and in a contrasting color the tee can be esthetically pleasing. Other materials that are used for ceiling tile are rigid cellulose lay-in tiles, ceramic faced sound-absorbent tile, and cement asbestos sound-absorbent tile. The latter are moisture and fire-resistant and can be used in exposed locations.

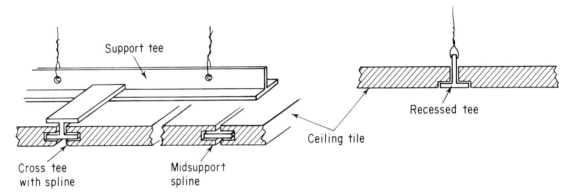

Hung Ceilings with Concealed Splines and with Recessed Exposed Tee.

2A1. *Composite Ceilings*

Several companies manufacture ceiling panels which perform the total job of a sound-absorbing ceiling; which supply and exhaust conditioned air, and which also contain built-in lighting fixtures. The supply air can be sent into the rooms by means of perforated holes in the tile or by means of slotted tees or by built-in diffusers. The exhaust air is returned through slotted tees which run into a return air duct. The architect engineer should take a long look at these combination ceilings and compare them in cost and appearance to an assembled ceiling in which the lighting fixtures, supply, and exhaust air can be put anywhere he wishes instead of being tied together in an assembly. Each system has its own advantages.

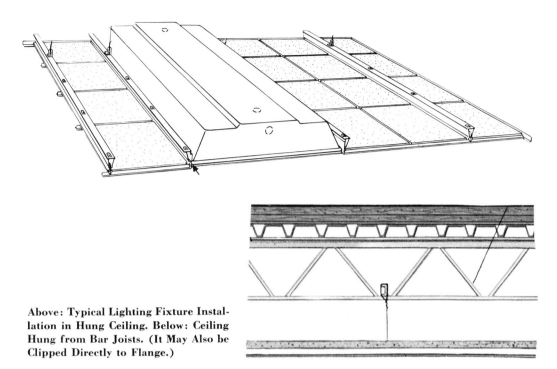

Above: Typical Lighting Fixture Installation in Hung Ceiling. Below: Ceiling Hung from Bar Joists. (It May Also be Clipped Directly to Flange.)

In smaller office buildings that use bar joists as floor supports the planner can use ceiling tile that can be fastened directly to the lower flanges of the joists and thereby use these flanges instead of hung tee supports. Such tiles can be metal faced and backed with sound absorbant materials such as glass fiber. The spaces in the web of these joists can be used to accommodate ducts, pipes, electrical lines; and also act as a return or supply plenum.

2A2. *Luminous Ceilings*

The completely luminous ceiling is often used by architect engineers in high-quality buildings. When done properly it provides a low-glare, low-concentrated light source which can be very handsome and gives a sense of spaciousness and luxury. There are many ways this ceiling can be erected. It should be remembered that the only breaks in such a ceiling can be for the actual support members. There can be no visible air supply or exhaust sources as such and, of course, the entire ceiling acts as a lighting fixture. The diagram shows one method of doing this. The main support tee holds slotted cross tees which are hollow and covered on *top and act as continuous supply or exhaust sources.* The supply tee is connected at intervals to a flexible supply duct and the exhaust tee is cross-connected to an exhaust duct. Care must be taken that the fluorescent tubes, which can be mounted on wiring channels, are far enough above the ceiling panel so that they do not show "hot spots" or identifiable sources of light. The ceiling panels are of acrylic plastic and can be flat, dished, or flanged to suit the decorative scheme. This

FLOORS AND CEILINGS

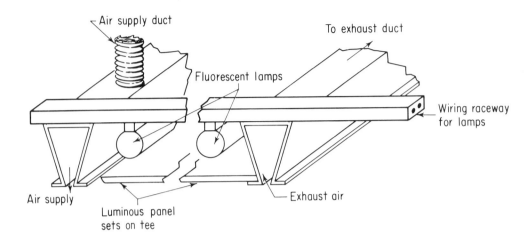

A Luminous Ceiling Installation.

is an expensive ceiling and must be carefully engineered to take care of special areas. It also should be carefully weighed before it is used. As usual the architect engineer and the owner must decide on the basis of a combination of esthetics and economics.

2A3. *Special-Use Ceilings*

In restaurant kitchens, and publicly used toilet rooms and shower rooms, the planner should use a sturdy, moisture proof, grease proof, sound-absorbent ceiling tile that can easily be kept clean. This tile usually has a perforated metal face backed with glass fiber or other sound-absorbent material. This backing should be carefully treated against mildew and should be permanently germ-proofed. A typical specification should call for a perforated aluminum panel not less than .032 inch thick, of a specified alloy, dead flat, and carefully squared for hairline joints. The perforations would be of ⅛-inch diameter on ½-inch centers in both directions. The backing of fiber glass should have a density equal to one pound per cubic foot and be of sufficient thickness to provide a noise reduction range of from 75 to 85%. The aluminum (or steel) should be given a baked-on matte white synthetic enamel finish.

In space such as fan rooms, compressor or boiler rooms, or elevator machinery rooms, where hung ceilings are not necessary but where noise transmission is a problem, the architect engineer can specify acoustical ceiling boards. These boards must be rigidly formed of fiber glass or other sound-absorbent material; painted with a white matte finish; rated incombustible; have a noise reduction coefficient of not less than 90% and capable of being spray painted at least eight times with nonbridging water base paint without lessening their noise reducing qualities. The usual thickness for such boards is 1 inch. They are held by short anchors (one per square foot of ceiling) to the floor above and a 1-inch space is left between the boards and ceiling. The boards can also be glued to the underside of the ceiling slab, but this is not as good a job.

2A4. *Acoustic Plastered Ceiling*

In most buildings there are spaces where tile ceiling should not be used. Banking space, lobbies, restaurants, or other places where the geometric formality of a tile ceiling is inappropriate, can use a hung ceiling finished with acoustic plaster. Instead of ceiling tile between the supporting tees the designer uses hung steel channels to which expanded metal lath is fastened. This metal lath is given a scratch and brown coat[2] and an acoustic plaster is troweled on over these gypsum plaster coats. It produces a ceiling without joints and textured in appearance. The lighting can be incandescent downlighting. It produces a pleasing and intimate appearance.

[2]This will be explained in Chapter 16, Section 3.

CONTENTS

1. A General Discussion of the Requirements
   A. The Character of Sound

2. Designing for Sound Conditioning
   A. The Private Home and Multiple-Family Dwelling
      1. A typical code requirement
   B. The Office Building
      1. Design for outside noise
      2. Design for inside noise
         a. Human noise
         b. Machine noise (airborne and vibration)
   C. Special Problems

14

SOUND CONDITIONING

1. A GENERAL DISCUSSION OF THE REQUIREMENTS

The architect engineer, when designing a structure in which people will live, or work, or play must keep in mind the total comfort of the inhabitants. He must provide them with enough light to see, and with fresh air, and with sanitary facilities. There is, however, another important comfort factor which must be considered and that is sound. In recent years, noise has become such a problem that it is frequently referred to as "noise pollution." In the private home or multifamily dwelling the transmission of sound from rooms and families can cause acute discomfort and annoyance. In an office, excessive noise or disturbing sounds can cause distraction and lead to a lowering of efficiency. Conversations overheard in private offices can cause embarrassment and sometimes serious trouble. The noise in a factory has on occasion actually caused deafness.

Playgoers and concert listeners are acutely aware of the sound qualities of a theatre or auditorium. Consequently, the sound conditioning of a space requires thorough knowledge of the materials and methods available for this purpose. This chapter will not attempt to go into the science of acoustics since this subject has received the attention of acknowledged experts for many years and a number of texts have been written about it. It will, however, give an owner or architect engineer a general view of the material and methods that are commercially available to produce comfort and convenience in the majority of situations that may be encountered.

SOUND CONDITIONING

1A. THE CHARACTER OF SOUND

Sound can be either wanted or unwanted. The sound of music in a concert hall, the voices of the participants in a conference, conversation in an office or living room is wanted. The noise of machinery, heavy traffic, or rushing air in a duct audible in an office or home is unwanted. Sound has an annoyance factor. The sound of one's wife banging pots around while preparing a meal may be pleasant, whereas the sound of your neighbor's wife doing the same thing may be downright annoying. The architect engineer must, therefore, in his design sort out wanted sounds from unwanted ones. He must attempt to create space in forms, shapes, and of proper materials so as to keep the wanted sound clearly audible without echo or reverberation and at the same time keep out unwanted sound as much as possible.

Sound has two basic qualities—intensity and frequency. The intensity is measured by a unit known as a *decibel* (db). The human ear can hear sound ranging from the "threshhold of audibility" to the "threshhold of pain." The sounds that can be heard range from 1 to 1 million million in what is known as the "middle frequencies." This vast range has been reduced to a scale ranging from 1 to 120 decibels. Sound is actually transmitted by air and exerts a measurable physical pressure. A decibel is therefore actually a measure of pressure and can be read on a meter that transforms this pressure to an electric current. The unit of pressure that is used is a dyne per square centimeter. (A dyne is the amount of force acting on 1 gram to produce an acceleration of 1 centimeter per second per second.) The decibel rating of sound intensity ranges from a whisper at about 15 decibels to ordinary conversation in an office at 55 decibels to 75 decibels in a heavily traveled street, and over 100 decibels in a boiler factory. Intense sound over a period of time causes physical fatigue.

Frequency is measured by cycles per second (cps) and the normal ear can hear frequencies ranging from 20 to 10,000 cps. The normal range of frequencies is from about 50 to 5000 cps. The human voice normally ranges from 128 to 1024 cps. As sound intensity or pressure is arbitrarily divided into units (decibels), so frequencies are also divided into units known as octave bands. An octave band is based on a geometric progression in which the high frequency in each band is twice that of the lower frequency. The starting frequency is at the lower level of audible frequencies such as 45 cps or 20 cps depending on which handbook or code one is studying and the octave bands are as follows:

$$\frac{45}{90} \quad \frac{90}{180} \quad \frac{180}{355} \quad \frac{710}{1400} \quad \frac{1400}{2800} \quad \frac{5600}{11200}$$

usually designated by mid frequencies at 63, 125, 250, 500, 1000, 2000, 4000 and 8000. *Or*

$$\frac{20}{75} \quad \frac{75}{150} \quad \frac{150}{300} \quad \frac{600}{1200} \quad \frac{1200}{2400} \quad \frac{2400}{4800} \quad \frac{4800}{9600}$$

All these ranges of sound intensity and frequency are available in tables that should be studied before the designer attacks the problems of sheltering them or blanking them out.

2. DESIGNING FOR SOUND CONDITIONING

In designing for sound conditioning the designer must know the level of sound that will be present in the space from outside sources, known as the *masking noise*. It may come from outside traffic, a nearby furnace, an electric motor, or the air conditioning system. There are allowable limits for such masking or background sound as, for example, 30 decibels in an auditorium or 35 decibels in an office. The acoustics engineer now uses another criterion for background or masking noise. This is known as the *noise criterion* (NC) which is derived from a standard curve of the maximum decibel (db) level in each octave band. This combination of frequency and loudness is what the ear receives. The architect engineer must first design the structure so that the noise made by motors, blowers, and ducts does not exceed the allowable limits. His second step is to design the actual space itself in order to keep the sounds within it audible and pleasant.

2A. The Private Home and Multiple-Family Dwelling

In the private residence the steps that can be taken to avoid annoyance sound transmission are relatively simple. The noise that may come from traffic, airplanes, or noisy neighbors can not really be controlled except by the use of double glass windows and glass fiber blanketing in walls and roof which produce a barrier to airborne noise. This blanketing acts as heat insulation as well. If the house is well built the normal masking or background noise should be about 20 db. The noises within the structure can be controlled in several ways. Airborne noise can be controlled by tight doors and substantial partitions such as metal lath and plaster instead of wall board. Resilient floor coverings or carpets reduce noise. The noise of an oil burner or blower motor can be reduced by isolating the machines on a cork or rubber base to take care of vibration transmission. If the house uses hot air heat the blower fan should be connected to the ducts by a flexible connection. If it uses hot water heat, the piping should be isolated from any structural members. The airborne noise can be controlled by plastering the walls and ceiling of the furnace room which should be done for fireproofing in any case. The heavier the materials that are used and the tighter the windows and doors, the quieter the house will be.

SOUND CONDITIONING

In the multiple-family dwelling, noise must be controlled by the use of positive isolation of airborne and solidborne noise. The problem of noise control has become so pervasive that it is now the subject of city codes and of national standards as promulgated by the Federal Housing Administration and the National Bureau of Standards. These regulations now force an owner to take the proper measures to control noise in residential buildings. It would be well for the architect engineer to become familiar with the requirements.

2A1. *A Typical Code Requirement*

A typical requirement will state the following: The design and construction of interior walls, partitions, floor-ceiling construction, and mechanical equipment shall be such as to provide minimum protection (as set forth), of each dwelling unit from extraneous noises emanating from other dwelling units and from mechanical equipment. In addition, airborne sound from exterior mechanical equipment in any building group must also conform to minimum standards. The regulation further states that maximum sound power from any mechanical equipment room must not exceed a certain amount (approximately 100 decibels at mid frequencies ranging from 63 to 8000 cps).[1] There are minimums set for the distance between any exterior equipment and any living quarters. There are minimums also set for the sound transmission class[2] (stc) of walls and ceilings surrounding mechanical equipment rooms (50 decibels) and between housing units and halls, or stairways. There must also be a minimum stc classification between housing units. This stc in all the above should be at least 45 decibels. Public authority has stepped into a situation which, had it been left alone could have jeopardized the health and welfare of the public.

2B. THE OFFICE BUILDING

The sound conditioning of a large or small office building has become the subject of intensive research on the part of the architect engineer and the acoustics expert. In the office "time is money" and inattention to the work at hand or annoyance caused by noise from within or without the work area, can result in significant loss of efficiency and even serious error. The sources of such noise must be identified and measures taken to eliminate them or at least reduce their intensity to a comfortable level.

---

[1]The architect engineer can design for these allowable ratings and test for them when the installation is complete.

[2]stc is the sound attenuation classification of the wall or ceiling. If a noise of 100 decibels occurs on one side, the wall will stop 50 decibels of it.

2B1. *Design for Outside Noise*

The first step is to take care of airborne noise that comes from outside the building. This can be done by soundproofing the curtain wall. With air conditioning now available the building can be practically sealed against outside air. The windows can be fixed in neoprene gaskets, or if the windows can be opened, they can be pivoted open and closed tightly on such gasketing. The metal portions of the wall should be backed with sound-absorbent material. In the case of masonry walls, they are either backed with lath and plaster or other furring material and generally have a very good attenuation rating.[3] Such construction reduces the average outside noise to no more than a background noise that is not annoying to the average person.

2B2. *Design for Inside Noise*

With the outside noise taken care of the architect engineer then investigates and provides controls for inside noise. This can come from typewriter clatter, general conversation, impact noise caused by high heels, partially heard conversation from the next office, the sound of air rushing through ducts, the hum and vibration of machinery, the vibration caused by pipes that carry pumped water, and many other sources.

2B2a. *Human noise*

The general noisiness of an office where many people work, converse, and use machines can be greatly reduced by the use of a sound-absorbent ceiling. As described in Chapter 13, such ceilings can be made of fiber glass, mineral fiber, or perforated metal backed with a sound-absorbent material. The ceiling is usually hung from a structural slab or floor above and its purpose is to receive noise and keep it. If the floor is of resilient tile, carpet, or cork, this also helps to absorb or prevent noise.

The normal office partition reaches only to the hung ceiling and there is therefore a space over it. This space can carry airborne sound. As stated before, the hung ceiling absorbs sound—but it is not monolithic. There are spaces around the lighting fixtures, ceiling diffusers, and exhaust grilles, where the ceiling abuts a wall; and all of these spaces, small as they might be, carry airborne sound from one office to another. The answer is to erect plenum barriers between office spaces or to build the partitions right up to the underside of the slab above. The latter is often impractical and costly. A plenum barrier is a sheet of material hung from the structural slab above and fastened to the top of the partition. It can be a rigid board, fiber glass, or thin lead. It stops most of the airborne sound over a partition. Impact

---

[3] Please see the footnote on page 240.

SOUND CONDITIONING

noise caused by hard heels, the moving of furniture, or dropped objects, can be reduced by resilient floors or carpeting and a hung ceiling below will generally absorb the rest.

2B2b. *Machine noise (airborne and vibration)*

We come now to the real design problem and that is the reduction of airborne noise and vibration noise from machines. The architect engineer starts at the source of the noise and progresses from there. The machine room itself must be soundproofed and removed as far as possible from work spaces. Machine room noise with compressors and pumps, etc. easily reaches 100 decibels. If it is in a basement area the walls surrounding it are usually of masonry and should be capable of absorbing at least 50 decibels. The ceiling can be covered with glued-on or hung acoustic tile or with a vermiculite acoustic plaster. All piping or ducts leaving the space should have fiber glass sleeves to close any openings. For machine rooms on roofs or in mid-building the precautions against airborne noise are the same. The rooms are enclosed in masonry and in the mid-building room the ceiling is soundproofed. Fan rooms are surrounded by louvred openings into the outer air and the fans and their heavy metal housings are soundproofed with fiber glass blanketing.

There are two other sources of noise from machines; one is the noise of air rushing through ducts on its way to cool or warm the building, and the other from the vibration of the machinery. All specifications call for sound attenuators in air ducts. Duct silencers are lining sleeves of fiber glass or other sound-absorbing material that are inserted into the ducts at points where they leave fan chambers, or at points where ducts enter a hung ceiling. High-velocity ducts should be located away from office space or carefully enclosed with heavy soundproof material.

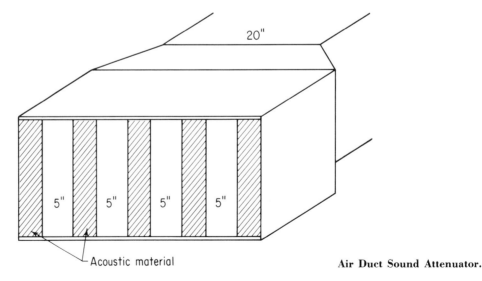

Air Duct Sound Attenuator.

SOUND CONDITIONING

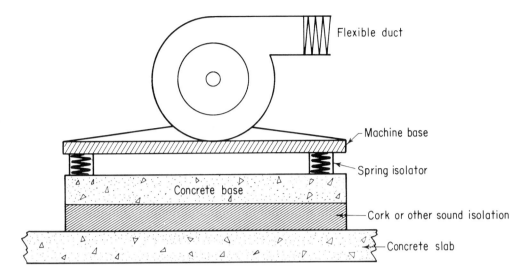

Vibration Isolation for a Blower or Other Machine.

Vibration noise can be stopped to some extent at the source by making sure that all machines are statically and dynamically balanced. When in use, however, fan blades will pick up airborne dust, and motor bearings will wear, and the balance is not likely to last very long. Care must therefore be taken that such vibration that does occur is not transmitted. The connection from a fan to a duct must be flexible as is the connection from any pump to a pipe. Even connections to electric motors should be made flexible. Then the machine should be mounted on springs or isolators over a concrete base, and the base in turn should be set on a pad of cork or other sound-absorbent material. The heavier the machine, the heavier the concrete block and springs. The concrete by its very mass or inertia resists vibration, and the springs and cork stop the rest. This is particularly important when the machine area is on a roof that is over the most valuable office space in the building. The architect engineer must also study the location of the machines with reference to the supporting structure. He should not, if possible, locate a machine in the middle of a span where it may set up sympathetic vibrations in the structure. If he has to, then he must thoroughly dampen its vibration. Certainly in large buildings using large machines the architect engineer must see to it that these machines when located on the roof or in mid-building over and above their own individual concrete bases, sit on heavy structural concrete slabs.

2C. Special Problems

There are, of course, special problems in noise transmissions, or noise abatement, or in sound conditioning which cannot be solved by normal means and require special research. Such sound conditioning problems would oc-

SOUND CONDITIONING

cur in the design and balancing of sound in an auditorium or concert hall. The author remembers a case where in order to test the acoustics of a new concert hall the architect engineer filled many seats with fibrous cones which had the same sound-absorbent and reflective character as people. Delicate balancing of acoustics requires that certain walls be sound-reflective or -absorbing and that distances and angles be carefully calculated. It is by no means an absolutely exact science. Other sound problems are more simple to solve. There are commercially available *"floating" floors* that are laid over sound- and vibration-absorbing material. Such floors will to a great extent blanket out from the space below, sharp "hammer" noises, or heavy foot traffic, or machinery noises. Where complete quiet between spaces is required there are *resiliently supported partitions* available. These in combination with plenum barriers, and fiber glass pipe, and duct sleeves give almost complete protection against airborne noise.

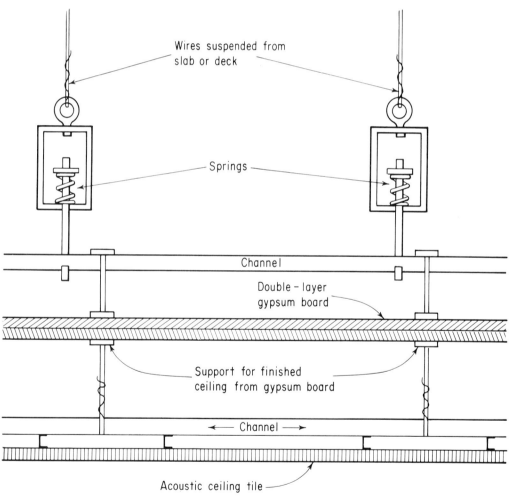

An Acoustic Suspended Ceiling.

Resiliently suspended ceilings are also available and the author knows of a case where the use of such ceilings satisfied a tenant and saved an owner from financial loss. In this case, although the main plant was in the basement, the fan room for the upper floors and several circulating pumps were located on the roof in an enclosed penthouse. Well before the building was completed, the upper floors were rented to a company that required large high-ceilinged soundproof areas. It was decided to locate the equipment on one side of the penthouse and to rent the other and larger portion to this company for this purpose. The agreement was that the landlord was to provide background noise environment for these spaces "typical of standard offices in good-quality buildings." Unfortunately, when these sound studios were completed and the fans, pumps, etc. were turned on, the sound pressure and frequencies exceeded the acceptable office standard of NC-25 to 35 and ran up to NC-60. Every piece of equipment was studied; fans were rebalanced; resilient springs were made heavier; all piping and ductwork were carefully isolated from every structural component and the building ducts were lined with sound mufflers. After each item was completed tests were made. Still too high! The final solution was reached by the use of two layers of ¾-inch-thick gypsum board hung resiliently from the slab with the finish ceiling hung below this. Expensive? Yes, but it need not have been if the problem had been recognized while the building was under construction. As in every other construction item, precognition of a problem can re-result in large savings in construction and rental dollars.

CONTENTS

1. STAIRWAYS—A GENERAL DISCUSSION
    A. Stairs in the Private Home
    B. Stairs in Fireproof Buildings
    C. Ornamental Stairways

2. ROOFING—AVAILABLE CHOICES
    A. The Private Home
        1. Flashing, leaders, and gutters
    B. The Small Office Building and Multiple Dwelling
        1. Roof decking and waterproofing
    C. The Institutional and Monumental Building
    D. Roofing and Flashing for Steel or Concrete Structures
    E. Factory and Warehouse Roofs
    F. Expansion Joints

3. MISCELLANEOUS IRON
    A. The Requirements
    B. The Specified Materials
    C. The Workmanship Requirements

15

STAIRS, MISCELLANEOUS IRON, AND ROOFING AND FLASHING

1. STAIRWAYS—A GENERAL DISCUSSION

Stairs are the basic means for the vertical movement of either people or material. No matter what other medium of vertical transportation is available, the stair is the only means that is accepted legally. The architect engineer can use the stairway as a focal point for an entire architectural scheme as witness the grand stairway of the Metropolitan Opera in Lincoln Center or the exterior stairways leading to the front plazas of office buildings or monumental buildings. Stairs can be the feature of the entrance hall of a private residence (we have all seen examples of beautiful curving stairways in fine homes), or the decorative stair in the entrance hall of a small special-purpose office building. In its usual form, however, the stairway is a wood, steel, or concrete structure which is utilitarian in purpose.

1A. Stairs in the Private Home

In the private home, when there is adequate space and funds, the designer can plan an entrance hall around a good stairway. In general, however, a house stairway is fairly utilitarian, with some decorative effect possible in handrails and balusters. Utilitarian or not, however, all stairs must be built to certain proportions of tread-to-riser, and wood stairs must be well built and wedged to prevent creaking or a feeling of instability. The tread-to-

STAIRS, MISCELLANEOUS IRON AND ROOFING

riser proportion can be obtained from a published table. The principle involved is: The shallower the riser, the deeper the tread. Most grand stairways have low risers and very wide treads. In a home, however, the architect engineer must carefully figure his proportion so that the stair meets the next floor and takes the least possible room for the most comfortable proportion (8" riser to 9'6" tread is fairly commonly used). He must also see that the stair is well glued, and that nailed and glued wedges are placed where riser and tread fit together and that the treads and risers are housed and glued into the string piece.

1B. Stairs in Fireproof Buildings

In any structure over two stories high housing multiple families or offices the stairway must be fireproof. The width of the stairs and the landings are determined by the population of the building and its use. All codes are specific about how such widths must be calculated, the sizes of stair doors, and their distances from the most remote part of the building.

In a structural steel framed building the stairs are always of steel construction. They are installed immediately following the steel so as to provide vertical transportation for workmen. The most common type of steel stair is the pan type. In this case the steel tread is set below the top of the riser so that the tread can later be covered with cement, usually with non-

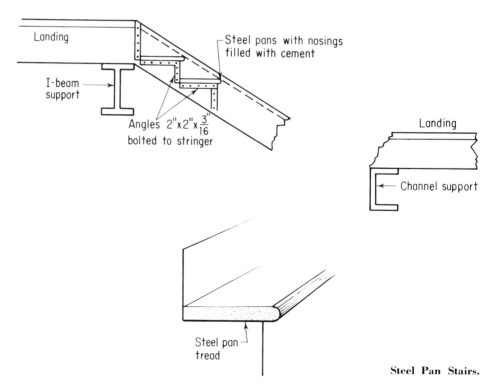

Steel Pan Stairs.

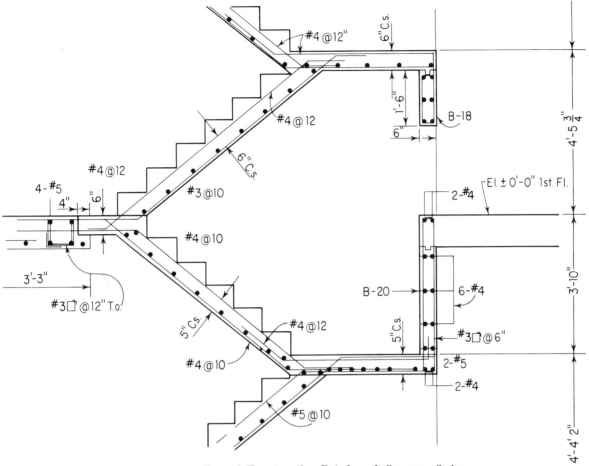

Typical Framing for Reinforced Concrete Stairs.

slip material added. The stairs are attached to the steel frame at each floor and the landings are hung from the steel above or supported on structural beams as shown in the diagram. Specifications for steel stairs emphasize proper fastening by bolting or welding, gauges of steel, and careful fitting to avoid projecting parts. Handrails should be smooth and at a convenient height, usually from 30 to 34 inches above the stair tread. The proportion of tread to riser for various types of occupancy is often a code requirement. Steel stairs are also built with the flat treads suitable as a base for precast terrazzo treads, slate or marble.

A reinforced concrete stair in a structural concrete building is usually poured just behind the structural slabs. The proportion of tread to riser is the same as for steel. Concrete stairs can be finished with precast terrazzo treads and risers or with a cement finish that contains nonslip ingredients. The stair is supported at each floor by reinforcing rods which have been left to protrude from the floor slabs and which are bent up or down to connect with the stair reinforcement.

STAIRS, MISCELLANEOUS IRON AND ROOFING

Curved Poured Concrete Stairway.

STAIRS, MISCELLANEOUS IRON AND ROOFING

1C. Ornamental Stairways

There are many excellent examples of ornamental stairways that are used by the architect engineer to add to the decorative scheme of an entrance lobby. There are open riser stairways with the treads supported in the center by a reinforced concrete stringer or steel channels and with treads of heavy reinforced precast concrete or terrazzo. Such heavy treads can also be supported at both ends by rods suspended from above. Such stairways can be curved or spiral. Shown here is an example of a handsome curved poured concrete stairway. The calculations for the proportion of riser-to-curve were done on a computer. Excellent stair design is limited only by the ingenuity of the designer.

2. ROOFING—AVAILABLE CHOICES

The architect engineer has a wide choice of roofing materials especially for private homes, smaller office and apartment buildings, and for institutional buildings such as schools, churches, and civic centers. The first duty of any roofing material is to keep out the elements. If it can also be sturdy, maintenance free, long lasting, and pleasing esthetically, it has fulfilled all requirements. The various materials that can be used have these qualities in different proportions. The architect engineer must choose the combination most suitable for the plan.

2A. The Private Home

The roof for a private residence depends on the climate, the architectural effect that is desired and the budget. The majority of middle-bracket homes with pitched roofs have wood or asphalt shingles. Wood shingles, usually of red cedar, come in different thicknesses and shapes and give a well groomed appearance or they can look quite rustic, as with hand split shakes. They are all installed in the same way. A series of wood furring or nailer strips (1" × 2") are fastened at right angles to the rafters at intervals determined by the amount of exposure each shingle is to get, i.e. 6 inches exposed to the weather, 7 inches and so on. The shingles are nailed to these strips, starting at the bottom and overlapping each row of shingles so as to cover the nails by 2 inches. Wood shingles usually overlap in three thicknesses. Red cedar shingles are resistant to weather and can last for many years without protective coating. It is recommended, however, that a coating of heavy-bodied shingle stain be applied or if little color is required then a transparent color stain may be used. Wood shingles can also be predipped in penetrating stain. If they are not coated they are likely to curl, especially when exposed to a great deal of sun. Asphalt shingles that are made of bitumen and fine grit come in many colors and several weights. The lowest weight is often used in development housing. Asphalt shingles are nailed over ⅞-inch-thick tongue and groove roof decking, or ½-inch plywood, either of which may be nailed to the roof rafters and then covered with an overlapping layer of tarred roofing felt or tar paper (15 pounds per 100 square feet). Heavier asphalt shingles are available. These are textured and have heavy butt ends for appearance and resistance to wind pressure, which in heavy storms is likely to tip up the edge of a lighter shingle. Asphalt shingles are fire-resistive.

Asbestos shingles are often used. They are made of asbestos fibers bonded in cement and come in many colors and textures. They are completely fireproof and are acceptable in city "fire zones" where noncombustible construction is required. Asbestos shingles are installed over tongue and groove roofing boards or plywood covered with roofing felt in the same way as asphalt shingles. Clay tile is used extensively in warm climates where a high degree of heat insulation is required. It is used in our own Southwest, and in California, and almost exclusively in Southern Europe. Clay tile comes in flat or curved shapes; it is practically indestructible and is expensive in this country. It usually comes in its own natural color, a warm reddish tan which weathers very well. It is laid over a board or plywood roof that is covered with tarred felt. Slate shingles are also used but they are quite expensive and their weight requires extra strength of framing. Both clay tile and slate require specially skilled workers who may be scarce in some areas.

2A1. *Flashing, Leaders, and Gutters*

The best roof that money can buy cannot keep water out of a house without the use of flashing. Flashing is the construction term for a metal or waterproof fabric that bridges the gap between abutting surfaces in different planes. It is used where there is a dormer or where a vent pipe or chimney comes through a roof. As the drawings show, the flashing interposes a sheet of metal to hold back moisture where a vertical meets a horizontal surface. The copper sheet is nailed against the vertical surface of the dormer wall and is then bent at right angles to lay flat against the adjoining roof. Both the vertical and horizontal surfaces are covered with the finish wall and roofing material. In the case of a plumbing vent a lead or copper cap is used. This bends over into the top of the vent and forms a collar around it at the roof surface. The collar is covered by the finished roof.

The valleys between roofs which meet at angles are usually flashed with copper, which fits in under the finished roofing. Copper is generally specified as 16 ounce cold rolled. It is usually wise to set the metal flashing in a bed of mastic, which retains its elasticity and serves as an extra barrier against moisture penetration. Flashing should also be used over window heads. The metal is nailed to the vertical surface above the window and bent out over the window frame to prevent moisture from driving into any crack between window

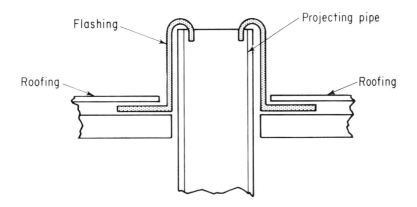

Flashing Of A Projecting Pipe

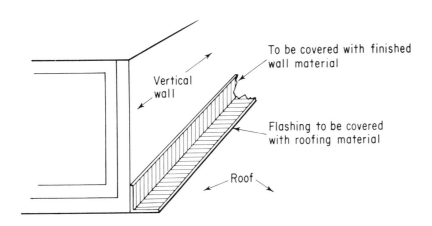

Flashing At Intersecting Surfaces

Flashings.

and wall. In houses with wooden gutters that are tight against the wall and the roof, it has been found that severe winter weather with alternate thawing and freezing can drive moisture up under the lower rows of shingles and cause roof leaks. A metal flashing going from gutter up to and under the first course of shingles can help prevent this.

Leaders and gutters for single-family houses can be of copper, although aluminum has become widely used. Wood gutters and copper or aluminum leaders are also extensively used. Gutters can also be concealed by inserting them in the roof, one course of shingles before the roof edge. These gutters can be connected to an internal leader to avoid unattractive leaders and gutters. This is expensive and is only used for high-cost residences or institutional buildings.

STAIRS, MISCELLANEOUS IRON AND ROOFING

2B. THE SMALL OFFICE BUILDING AND MULTIPLE DWELLING

2B1. *Roof Decking and Waterproofing*

The architect engineer can consider the roof decking and waterproofing as a single entity in a small business building or apartment house. Usually the roof is flat and there is a choice of many kinds of roof deck or slab. The roof deck may or may not be the same material as the floor system and the roof sandwich or composite must have two qualities that the floor system need not have: One is a vapor barrier and the other is heat insulation value. Roof decking can be of cellular steel laid on steel bar joists. The steel decking can be covered with insulating board or it can be covered with acoustic material underneath. There is an acoustic decking that contains holes in the sides of the corrugations and is packed with acoustic material such as glass fiber. Over this is laid heat insulating board. The underside can serve as a ceiling for certain spaces. Instead of insulating board the planner can cover this cellular roofing with lightweight insulating concrete containing perlite[1] or vermiculite.[2] Each of these may be obtained as precast slabs which can be laid directly over roof beams. There are several varieties of precast concrete roof slabs available. These are reinforced with steel mesh and are laid over and fastened directly to the roof beams. They also come with steel edges or with tongue and groove edges for close fitting. For a longer span, slabs of cellular formed concrete, reinforced with rods and mesh, may be used. Such a slab, 4 inches thick can carry 40 pounds per square foot over a span of 18 feet. Some slabs combine insulating and structural qualities, such as the asbestos faced roofing board containing a wood fiber filler bonded under high pressure.This 2-inch-thick board can carry 131 pounds per square foot in a six-foot span. There is a roof decking material consisting of chemically treated wood fiber bonded with Portland cement. This has excellent acoustic and heating insulating qualities. This deck can be cut and fitted by carbide tipped woodworking tools. Finally there is precast or poured-in-place gypsum roof. The poured roof is laid over form boards that are permeable to moisture and air and which are fastened to roof beams. The form boards are covered with a steel reinforcing mat and then 2 inches of gypsum fiber concrete containing wood chips is poured over this mat. The gypsum sets almost immediately and waterproof roofing can be applied within 20 minutes. The slab dries from underneath through the form board and the inner surface must be carefully ventilated at all times.

In all cases where there is a high indoor relative humidity or where there is any question of remaining construction moisture, it is recommended that a continuous impermeable vapor barrier be applied to the roof slab be-

---

[1] Perlite is a volcanic rock that expands or pops when subjected to intense heat. Its resultant trapped air cells have high insulating value.
[2] Vermiculite is a mica material that is treated and acts the same as perlite. Both materials can be used in insulating plasters.

fore any insulating board is applied. Insulating board can be asphalt impregnated fiberboard, rigid cellular glass, or compressed cork. Because of their fibrous or cellular construction with the consequent entrapped air, these boards have high insulating value. For instance, on a 4-inch concrete roof slab, a 3-inch insulating board will reduce the "U" Coefficient of Transmission[3] from .70 to .10. In combining a vapor barrier and insulating board the architect engineer should remember that the insulating board must be thick enough to maintain a temperature above the dew point of the indoor air in cold weather or air vice versa in warm weather especially if the interior is air conditioned. Failure to do so will cause condensation on the ceiling below or under the insulating board.

All the various forms of decking mentioned here have to be waterproofed. The usual form of waterproofing, especially for a flat roof, is by means of a number of layers of tar-impregnated felt paper of various weights, and hot pitch. Many architect engineers specify a "Bonded Roof." This roof must be laid in strict accordance with the manufacturer's specifications. Upon completion and inspection a ten- or twenty-year bond is issued which guarantees to keep the roof watertight for that length of time (except for accidents or extraordinary use). Aside from the specifications for the actual roofing, the requirements for a bonded roof on a roof deck are as follows.:

> All roof decks must be securely fastened structurally to prevent any slippage or movement.
> Steel deck must not deflect more than ⅛ inch at the center of a span and must be covered with at least 1-inch-thick insulation board.
> Wood deck cannot be less than 1-inch thick on rafters not exceeding 24-inch O. C. Any knot holes or other large imperfections must be covered with sheet metal.
> Precast concrete slabs must have all joints caulked or pointed with asphalt cement.

The composition of the roof itself is explained in Section 2D of this chapter.

2C. THE INSTITUTIONAL AND MONUMENTAL BUILDING

We now come to the roofing of a noncommercial building in which the designer wishes to convey an air of permanence. Two of the most commonly used materials for pitched roofs for such structures are copper and a copper bearing steel that is coated with a lead-tin alloy called *terne*. Copper is one of the most enduring of structurally used metals. Because copper has a comparatively high coefficient of expansion, the architect engineer must be sure that the copper specified is of the correct hardness and weight to resist

---

[3] The Coefficient of Transmission "U" is the Btu per square foot transmitted per degree Fahrenheit difference in temperature between the two surfaces of a material.

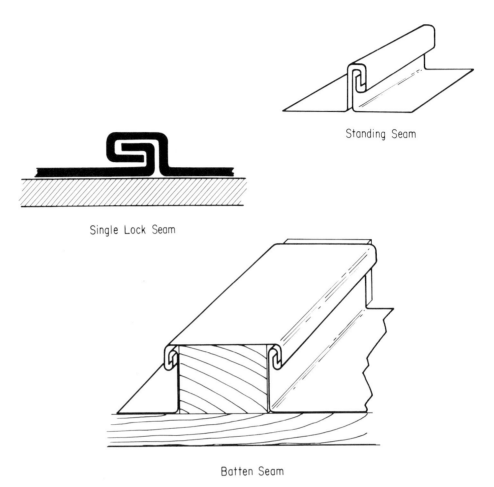

Types of Metal Roof Seams.

buckling. Three of the numerous ways in which copper can be applied to show various patterns are shown here. Copper weathers to a pleasing color and lasts indefinitely. Terne metal can be applied in the same way as copper. It can be left in its natural color or painted. Terne metal has high tensile strength and a low coefficient of expansion.

2D. ROOFING AND FLASHING FOR STEEL OR CONCRETE STRUCTURES

The steel or concrete structure for a large office or apartment building is flat roofed and the roof slab is poured, reinforced concrete. The roofing is therefore confined to the applied waterproofing, insulation, and flashing of this roof. The best way to describe the finished roof in this case is to state the requirements of a typical specification: the work included is a *twenty-year bonded,* insulated, and gravel surfaced 4-ply insulated roof; copper flashing as shown including compression seals around piping, ducts, or structural members protruding through the roof surface complete with flashing collars; pitch pockets; copper gravel stops where shown; flashing around roof scuttles and skylights.

The materials to be used are as follows: the "primer" and "coal-tar pitch," both of which must conform to specifications of the ASTM; the coal-tar saturated felt to weigh 15 pounds per 100 square feet and to conform to ASTM standards. The rigid insulation board to be of cellular glass, or pressed fiber, or compressed cork which must have a very low factor of moisture absorption and must conform to a Federal Specification of Insulation. Copper for flashing, which should be lead-coated with the lead weighing at least 6 pounds per 100 square feet on each side. The cap flashing should be at least 16 ounces and the base flashing must be 20 ounces; all copper should conform to the standard specifications for Copper Sheet and Lead Coated Copper Sheet of the ASTM. The sealant material should be snythetic rubber (polysulfide liquid polymer-thiokol base), and should be suitable for application with a caulking gun. The nails, screws, or bolts and any other metal accessories should be of copper or bronze.

No roofing operation should start until the underlying surface is thoroughly dry and is completely clean. Roofing should not normally be laid at a temperature under 50°F. The roof surface should first be mopped with hot primer (which is a coal tar) and hot coal-tar pitch. The roof insulation blocks are then laid into the hot pitch and pressed firmly into it. The insulation is then covered with four layers of tar-saturated rag felt, each layer laid in a bed of hot pitch and so overlapping the layer beneath that there are four layers of felt and hot pitch over the entire roof surface. The top layer of hot tar should have roofing gravel embedded into it to form a hard walking surface.

If roofing has to be laid in under 50° weather, it must be done carefully to avoid buckling later. The under surface must be thoroughly dry and the mopped hot tar must be immediately covered with the roofing felt. All the insulation and all the plys of roofing must be done in quick succession to avoid excessive cooling of the tar. When the work stops it must be quickly sealed off against weather by a "cut off" which extends the waterproofing at least 4 inches beyond the completed roofing.

Copper flashing generally comes in 8-foot strips that are soldered together with an expansion joint every 24 feet. The cap flashing is fastened into the vertical surface and caulked with a sealant. The base flashing is laid over several layers of felt, which are turned up under the cap flashing. The base flashing is cleated to a wood strip imbedded in the tar and then folded over and up under the cap flashing. The horizontal leg of the base flashing is then covered with two more layers of felt imbedded in hot tar.

Copper flashing is also used around vent lines, or flagpoles, or other round objects protruding through a roof. The pipe or pole is surrounded by a metal ring which holds a metal hood tightly to the pipe. This metal hood acts

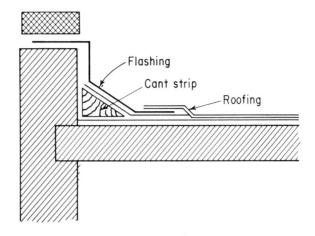

Cap Flashing

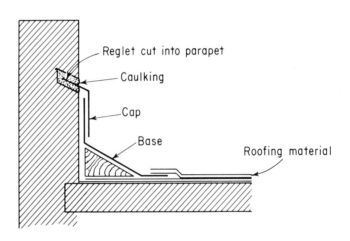

Cap And Base Flashing

as the cap flashing. The base flashing is installed around the pipe and soldered in place. It is then covered with layers of hot pitch and felt. In cases where structural steel members of guy wires come through, roof waterproofing is usually done by pitch pockets. These are copper boxes open at the top and flattened to lie horizontally at their bottom edge. The horizontal portion is covered by waterproof roofing. The open boxes are filled with pitch which seals the space around the structure. Gravel stops are raised metal strips placed at the edges of roofs to stop imbedded gravel from being washed over the roof edge.

STAIRS, MISCELLANEOUS IRON AND ROOFING

All this roofing work must be carefully done and throughly inspected. A leaking roof in a large structure is a special kind of nuisance and the leak may be very difficult to locate.

2E. FACTORY AND WAREHOUSE ROOFS

Roofing for factories and warehouses with pitched roofs can be done very simply. There are a number of metals that can be used. They come in almost any desired length and are easy to install. Aluminum corrugated roofing comes in several thicknesses from .032 to .050 inch. It will support 60 pounds per square foot on 8'6" spans for the .050-inch thickness. Aluminum is corrosion-resistant and very light for handling purposes. It is nonmagnetic and nonsparking and can be obtained with color plastic coating.

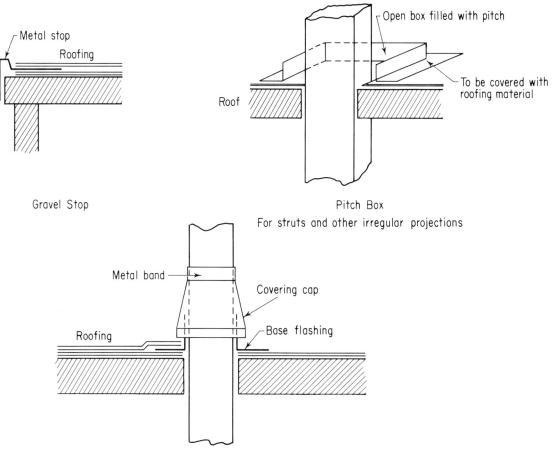

STAIRS, MISCELLANEOUS IRON AND ROOFING

Stainless steel corrugated roofing has high tensile strength and great resistance to heat. It is corrosion-resistant and comes in light gauge for easy handling. Stainless steel has a lower coefficient of expansion than aluminum.

2F. Expansion Joints

All roofs, no matter of what material, must have allowance for expansion. The expansion joint should allow free movement of the roofing material, and at the same time keep the roof weather tight. There are many variations of an expansion joint as shown in the diagrams. There are also expansion devices used for flashing. Expansion joints are also used on other surfaces that must remain weather tight such as sidewalks or plazas over subgrade structures. These usually consist of brass or stainless steel ribbed

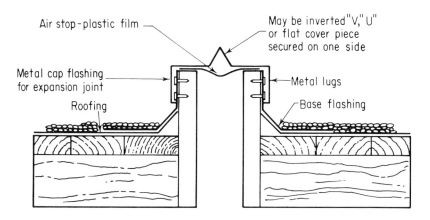

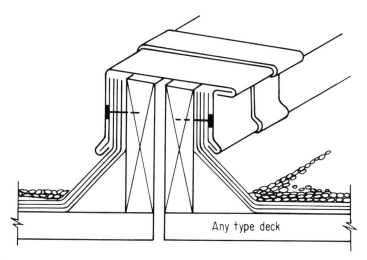

Expansion Joints.

STAIRS, MISCELLANEOUS IRON AND ROOFING

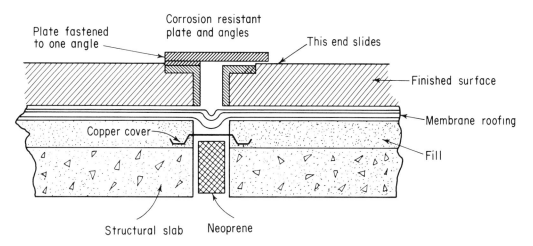

A Plate Expansion Joint Used in Large Open Areas.

plates that are fastened to one side of the joint and allowed to slide freely over the other side. The waterproofing of the joint which is under the plate device also has provision for expansion.

3. MISCELLANEOUS IRON

The miscellaneous iron work in any building is usually a catch-all for the various types of metal fittings and accessories that are required in a building but which are not part of a particular trade subdivision. They may be individually small but they are important and must not be forgotten.

3A. The Requirements

The requirements for miscellaneous iron includes labor and material to install the following items:

> Inclined and vertical ladders and open riser type stairs complete with platforms, treads, rungs, and railings.
> Interior and exterior steel pipe railings where not used for stairways.
> Interior and exterior gratings including grating floors and catwalks for mechanical shafts or smoke shafts, complete with framing struts, bracing, and hardware as required.
> Loose steel lintels.
> Hatchway and trap doors in floors and ceilings with framing and hardware.
> Trench and pit covers of ferrous metal with framing and hardware.
> Curb angles, column guards at loading docks.
> Tracks for window washing machines.
> Access doors with framing and hardware.
> Raised pattern safety steel plate platforms with framing and hardware.

STAIRS, MISCELLANEOUS IRON AND ROOFING

Projected steel sash in roof bulkheads with glazing and hardware.
Steel plates and angles for parapet masonry.
Structural framing for steel roll-up doors.
Abrasive cast iron stair nosings.
All rolled shapes, bent plates, sleeves, clips, brackets, hangers, anchors, bolts, and other fastenings and supports required for the job but not specifically mentioned as part of the work of any other trade.
The scrupulous specification also mentions "Work not Included." This is for the protection of the owner and the contractor who might otherwise bid a very high price to cover him in any contingency.

3B. THE SPECIFIED MATERIALS

The quality of the materials to be used is tied to standard specifications by the architect engineer. The structural steel, steel sheets, steel plates, wrought iron, steel pipe, gray iron castings, malleable iron castings, abrasive castings, hot dip galvanizing, etc. are all specified to conform to ASTM standards, or Federal standards for these materials. Where necessary the required tensile strength is mentioned.

3C. THE WORKMANSHIP REQUIREMENTS

The workmanship required for miscellaneous iron is carefully spelled out. All the items that are built or installed in this category are made of castings and rough iron shapes and they must be assembled properly to prevent stumbling, and cuts and abrasions to the people who will be walking over them, passing them, or handling them. Most of this installation is never seen by the public but is used constantly by workmen. Welds must be ground smooth. All exposed joints must be close fitting. All bolts and screws must be cut off flush with adjacent metal. Threaded connections must be made tight so that no sharp threads are exposed. Metal work must be fastened to masonry with Rawl plugs and to hollow masonry with toggle bolts. In addition, this division provides work platforms which sometimes must bear heavy loads. All ladders or inclined open riser stairs must safely support at least 100 pounds per square foot. All grating treads should have cast iron abrasive nosings. Ladder rungs should be of galvanized wrought iron. Gratings must be designed to support a live load of 300 pounds per square foot. (The reason for this high figure is that such gratings are usually used when repairs are to be made and they must carry workmen, tools, and material.) Gratings have to be carefully framed around ducts and pipes so that the strength of the grating is not reduced. Steel plates and walkways must be carefully fitted and strong enough to support at least 100 pounds per square foot. Hatchway doors must be of raised pattern safety plate and must be flush with the finished

STAIRS, MISCELLANEOUS IRON AND ROOFING

floor. Corner plates and edge plates have to be installed with no projecting edges and projected steel sash should work freely and be accurately balanced when opened.

The architect engineer should study his plan carefully before specifying this section of the work to make sure he has included everything necessary to enable the building to perform its proper functions.

CONTENTS

1. INTERIOR MASONRY—WHERE IT SHOULD BE USED
    A. Code Requirements
    B. Available Material
    C. A Typical Specification

2. LATHING
    A. Expanded Metal Lath
    B. Gypsum Lath

3. PLASTERING
    A. Gypsum Plaster
    B. Keene's Cement Plaster
    C. Vermiculite Plaster
    D. Acoustical Plaster
    E. Portland Cement Plaster
    F. Workmanship Specifications

4. PAINTING AND DECORATING
    A. Available Coating Materials
    B. Workmanship Specifications

16

INTERIOR MASONRY, LATHING, PLASTERING, PAINTING, AND DECORATING

1. INTERIOR MASONRY—WHERE IT SHOULD BE USED

Interior masonry in any structure built for normal business or residential use is generally confined to partitions between living units in multiple dwellings, between separate tenants in business buildings; and to enclosures around fire exits, stairways, elevator shafts, or hallways. Masonry is also used to fireproof steel columns, wind bracing, struts, and other structural members. The use for interior masonry is mostly for fire protection. In order to determine the requirements for interior masonry the architect engineer must consult the local building code having jurisdiction, either city, state, or national. All codes classify structures by use, height, area, and by the district in which they are located. Each classification of structure gives a required fire-resistance rating in hours for all its component parts.

1A. Code Requirements

Some of the requirements are as follows: Hung ceilings contributing to the required fire-resistance must be fire stopped for every 3000 square feet.[1] (While this need not be done by means of masonry, it is wise to know

[1] This area may differ by locality.

BUILDING & INTERIOR

about it.) Penthouses or other roof structures must be built of masonry or other noncombustible material.

The fire divisions between different occupancies in business or residential buildings beyond certain sizes require a high fire-resistance rating, the requirements of which can often best be satisfied by masonry. This is also true for all shaftways which require the maximum hourly rating for the particular structural type and masonry is the best means to accomplish this.

1B. AVAILABLE MATERIAL

Unless there is a decorative reason for doing otherwise, interior masonry units are available in gypsum, terra cotta, and cast concrete that can be plastered on one or both sides. The terra-cotta or clay block and the cast concrete block can be obtained with a decorative facing in the form of a fired on glaze. The gypsum block may be left as is in unfinished areas and is usually plastered in finished areas.

Gypsum block comes in three standard thicknesses—3-inch, 4-inch and 6-inch. Depending on whether it is plastered on one or both sides and on the thickness of the plaster, gypsum block can be used in walls of up to 4-hour rating. It is laid in a gypsum type mortar and is used quite extensively in demising walls between tenants, for corridor walls, and for fireproofing around structural steel columns.

Clay tile is hard-burned structural quality clay and can be used in the form of 2, 4, 6 or 8-inch thick hollow terra-cotta blocks for the same purposes as gypsum, and for stairway and shaft fireproofing. As a rough block, clay tile comes with a smooth or a scored face so that it can be plastered or left as is in unfinished areas. A great deal of surface glazed structural clay tile is used for finished surfaces in hallways, toilet rooms, and stairways. The glazed surface comes in a variety of decorative colors. The tile is abrasion-resistant, easy to keep clean, and resistant to staining or soiling. It is available in many shapes such as rounded corners, coved base, curved sills and jambs, and bullnosed ends. Clay tile is also available with acoustical qualities.

Concrete block is a casting of Portland cement and lightweight aggregate and comes in thicknesses from 2 to 12 inches. It can be obtained as a solid block in the thinner versions but is almost always hollow in the larger sizes because it would be too heavy to handle if it were solid. It can be obtained with one or two color glazed faces or as a plain faced gray colored block. Like clay tile it can be obtained in various shapes. Concrete block in its natural color is used widely as an enclosure for stairhalls and shaftways. It can be plastered, and painted or wallpapered, or faced with marble on the finished side, and presents a uniform finely-grained neutral-colored surface on the unfinished side. With the use of lightweight aggregate it is comparatively easy to handle.

1C. A Typical Specification

A typical specification for concrete block is as follows: The concrete block or masonry unit should be made from Portland cement and lightweight aggregates. After casting, the unit should be allowed to set for not less than 3 hours and then is to be steam cured. Steam curing shall be for not less than 8 hours under a steam pressure of between 120 and 150 psi and a temperature of between 340° and 370°F. The block should be laid in a mortar of Portland cement, sand, and lime putty. The block should be solidly bedded in mortar with the vertical joints well buttered and not more than ⅜ inch wide. The vertical joints should break half way over the course below. (For decorative purposes the block is sometimes laid without breaking courses.) All masonry walls must be laid full height so as to terminate at the underside of the structural slab above, and must be anchored to the structure by metal anchors. Masonry partitions should be carefully cut to fit around ducts and pipes. All necessary cuts in masonry units should be made by power driven carbide saws.

A well laid concrete block partition with even joints and true vertical and horizontal lines can be quite pleasing in a stairhall or in a classroom. The surface can be painted in many colors for excellent decorative effect.

2. LATHING

"Lathing" as a construction term denotes the covering of structural members by a material that is used as a base for the application of a finishing material. The lath and the finishing material form a surface that conceals these members. The construction process involved is almost always referred to as *furring and lathing*. The furring strip is the metal, wood strip, or masonry unit that is applied to a wall or a ceiling to form an insulating air space and the lath is applied to this furring material. The lath can, of course, also be applied directly to the structural member or wall.

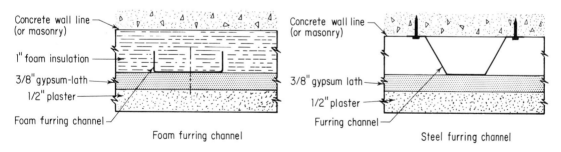

Furring Sections. Furring Can be of Wood.

2A. Expanded Metal Lath

The two materials most widely used for lathing are expanded metal lath and gypsum lath. Expanded metal lath is formed from various gauges of sheet steel which are punched into a series of small diamonds that look not unlike a honeycomb except that sharp edges project to form a bond with plaster. Metal lath comes in many other shapes and in a number of weights. It always has sharp edges to bond to plaster or cement.

In higher-class private residences, walls and ceilings are still finished with metal lath and plaster. The lath can be nailed directly to the underside of floor joists and to wood studs. It forms an excellent base for plaster and is valuable for its fire-retardant quality. Metal lath is used extensively as a base for plaster behind bathroom tile walls or as a base for cement plaster walls between an attached garage and the main house as called for by most codes. It is also used around furnace rooms or other areas with any special fire hazard.

Metal lath is also used in hung ceilings where it is wired to steel channels which are suspended from the underside of the floor above. The lath should be installed so that the joints between the sections are staggered to avoid cracking. The architect engineer should also call for expansion joints not more than 30 feet apart in any direction. Because metal lath is a flexible material and can be shaped easily, it can be used to form cornices, ornamental moldings, or arched or domed ceilings. It can also be used for exterior work. In this case a heavy gauge of lath is used. The lath is wired over steel channels and then covered with a concrete mixture such as "gunite," a small aggregate concrete which is forced from a nozzle under high pressure. Self-furring metal lath can also be obtained. It contains small protuberances that hold it away from a wall to form an air space. Self-furring metal

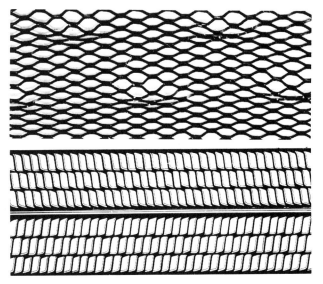
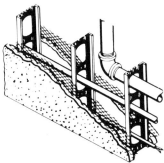

Left: Typical Metal Lath. Right: Section of Heavy Channel for Bathroom (etc.) Wall. Below: Typical Metal Rib Lath.

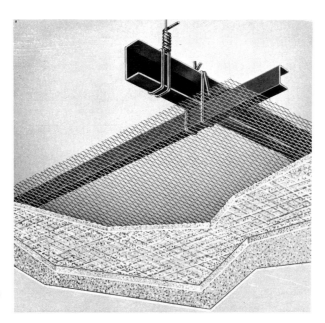

Hung Ceiling of Metal Lath and Plaster.

lath can also be obtained as expanded rib lath, which has protruding solid metal ribs between the perforated portions to form structural stiffening, as well as forming an air space between lath and wall.

Thin (2-inch) fire-retardant partitions are usually made of metal lath that is wired to steel channels and then plastered solidly on both sides. Metal lath is used to fireproof steel, in which case it is wrapped around steel columns or under and around steel beams, and is covered with several coats of plaster. Even when other lathing material is used, the elasticity and strength of metal lath makes its use necessary for expansion joints, corner beads, cornerite, or over doors to prevent plaster cracking. The corner bead presents a rounded metal edge at corners to prevent chipping, as the diagram shows. The cornerite is used in inside corners to prevent cracking.

A typical specification for metal furring and lathing in an office building should contain the following provisions:

> The work should include the furnishing and installation of hangers for suspended lath and plaster ceilings; the furnishing and installation of hangers and lath for plastering under steel stairs; the metal furring and lathing for vermiculite or perlite plaster fireproofing; and for all finish plaster walls and ceilings including solid plaster partitions; the furnishing and installation of corner beads, expansion joints, metal lath reinforcement at corners of openings in masonry surface or at the junction of dissimilar surfaces to be plastered.

The work must conform to various codes having jurisdiction, and all work and materials must conform to standards of quality set by the American Standards Specifications.

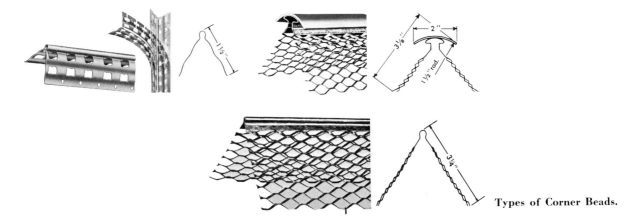

Types of Corner Beads.

The quality of the material should be as follows: All metal lath must be either galvanized and coated with rust-inhibitive paint or be of stainless steel. Corner beads and expansion joints must be of stainless steel. The plain lath should be of diamond mesh pattern formed from sheet steel and should weigh not less than 3.4 pounds per square yard. The flat rib lath should weigh not less than 3.4 pounds per square foot; the ⅜-inch rib lath 4 pounds and the ¾-inch rib lath 6.75 pounds. (The latter laths are used as self-furring lath and are used for long unsupported spans, or where a heavy weight of plaster is necessary for fireproofing.) The quality of the workmanship is specified by certain provisions such as directing that the lath be laid with its long dimension at right angles to the furring channels and specifying a wire tie to the channels at not less than 6-inch intervals. The lath should lap at least an inch at sides and ends and laps should occur only at supports. For chases (furrows or holes cut into a masonry surface to conceal piping, etc.) that have to be covered with lath it must lap at least 3 inches on each side and if the chase is more than 12 inches wide it must be bridged with furring channels on 12-inch centers and lap 3 inches on either side, before being covered with plaster.

If the specification is studied it will be evident that its purpose is to provide a sturdy, rustproof, crack-resistant base for plaster. It is much easier and cheaper to do this at first than to come back later to repair cracks or sags in ceilings or walls.

2B. Gypsum Lath

Gypsum lath is used extensively in private residential work. It can be obtained in different thicknesses and in various forms such as foil-backed for insulating exterior walls or smooth-faced or perforated for the keying-in of the plaster.

Gypsum lath when used in the private residence should be firmly nailed at each stud and each lath must end on a stud or joist. A ⅜-inch-thick lath with a ½-inch plaster coat over it has a fire rating of 1 hour which is all

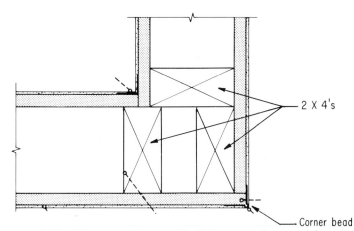

Construction of an Inside and Outside Corner Using Wall Board.

that is required in a normal sized house. Gypsum lath is manufactured in a waterproof form for use in bathrooms and shower rooms behind tile. It can be used in a 2-inch fire-retardent partition by being wired to vertical ¾-inch steel channels and then being covered with three coats of plaster to form a solid wall. Gypsum lath can be used to fireproof columns and in combination with various thicknesses of plaster can be rated up to 4 hours.

Gypsum wallboard is used very often instead of a combination of lath and plaster. It is used most frequently in development housing. It comes in sheets 4′ × 8′ and in various thicknesses. It is made of a sheet of gypsum which is covered on both sides by a fiber-like paper.

Wallboard is, of course, cheaper than the combination of lath and plaster and has the advantage that it can be painted or papered almost immediately after it is installed without the drying period necessary for plaster. It is not as sturdy as lath and plaster and does not have the same fire-retardent qualities. If wallboard must be used the architect engineer should take some simple precautions to see that it is used properly. It should be throughly nailed at every stud and joist. The corners should be firmly butted over a sturdy corner post as shown. The joints should be taped in not less than three operations and carefully sanded each time, allowing several days of drying time between each operation and before it is painted or papered. A board thickness of ⅝ inch is recommended.

3. PLASTERING

3A. Gypsum Plaster

Plaster generally consists of a mixture of gypsum plaster and sand. It is applied in three coats on metal lath. The first, or scratch, coat is mixed in the proportion of one part of gypsum neat plaster[2] (fibered) to two parts

---

[2] Gypsum neat plaster consists of not less than 60.5% of calcined gypsum (calcium sulphate) plus other material to control setting time. The fiber is a hair fiber used as a binder.

of sand. The second, or brown, coat is mixed in the proportion of one part gypsum neat plaster (fibered) to three parts of sand. The third, or finish, white coat is mixed in the proportion of three parts of lime putty to one part of gauging plaster.

Lime putty consists of hydrated lime that has been slaked with water and mixed to a stiff consistency and then soaked in water for 24 hours. Gauging plaster is a quick setting plaster like plaster of paris. Once the gauging plaster is mixed it must be used immediately.

Gypsum plaster is applied in only two coats (brown and finish) on masonry and in one white finish coat over a bonding agent to a concrete ceiling surface. Where gypsum plaster is used in two or more coats the plaster should be at least ¾-inch thick. It finishes with a hard glossy white coat ready for decorating.

3B. Keene's Cement Plaster

Keene's cement finish plaster consists of the same mixtures of gypsum plaster for the necessary undercoats and a finish coat of two parts of Keene's cement to one part of lime putty. This finish coating forms a hard dense abrasion-resistant, moisture-resistant, impermeable surface. It is used where the ordinary white finish coat would crumble and disintegrate.

3C. Vermiculite Plaster

Vermiculite plaster consists of the same mixtures as gypsum plaster, with the addition of vermiculite (a mica material expanded under great heat) in the proportion of 100 pounds of gypsum neat plaster to 2½ cubic feet of vermiculite. It is applied in the same number of coats as the other plasters. Vermiculite plaster is used for fireproofing purposes in boiler room ceilings, around steel columns and beams, and in high hazard areas. The thickness of this total coating depends on the fire rating required by the local code.

3D. Acoustical Plaster

Acoustical plaster is a mill-mixed product which varies in its ingredients. It is usually sprayed on in two coats over a base coat of ordinary gypsum plaster. It can be trowled if required. The recommended thickness of the acoustic plaster is ½ inch. The brands vary somewhat in their acoustical quality and light reflectance. The specification, however, always calls for this plaster to keep its acoustical quality after being painted (future maintenance) at least four times with a nonbridging water base paint.

3E. PORTLAND CEMENT PLASTER

Portland cement plaster consists of one part of Portland cement to three parts of sand, by volume, to which is added lime putty in a volume of 25% of the cement volume. It is applied in two or three coats, the same as the gypsum plasters, with a minimum thickness of 1 inch. Portland cement plaster is used for the exterior stuccoing or coating of masonry walls. For interior or exterior work it presents a hard, durable, abrasion-resistant and waterproof surface which will give a top fire rating. It is used as a scratch coat behind wall tile in toilet and shower rooms and in other places of high moisture content.

3F. WORKMANSHIP SPECIFICATIONS

The workmanship specification for all these plasters is somewhat the same. It calls for the scratch coat on metal lath to be applied with sufficient pressure to key into the lath and to be applied in a thickness equal to one-half the total thickness required. The scratch coat is to be cross-scratched with a trowel to bring it to a rough texture. The brown coat should be applied when the scratch coat has hardened but is not dry. The brown coat should be brought to a uniform level finish flush with the grounds (plaster grounds are wood strips fastened at intervals to the surface to be plastered and which serve as guides for the finishing thickness and leveling of the plaster). The brown coat must be allowed to dry thoroughly before the finish white coat is applied. If the finish coat is Keene's cement, the brown need be only partially dry. There is another finish known as *sand float finish*. This is a mixture of white sand and gypsum plaster. It is brought to a fine grained finish by "floating" the moisture and fine aggregate in the plaster to the surface by rubbing with a wooden trowel. It makes a very pleasingly textured surface.

The workmanship specifications are explicit about working temperatures. No plastering can be allowed unless the structure has been kept at a temperature of not less than 55° for a week before, and for a week thereafter. The spaces must be well ventilated to promote even and not too rapid drying. There are other plaster mixtures such as "perlite," which is an expanded volcanic stone used in acoustical and fire-retardent mixtures, and other patented special-use mixtures, but the ones mentioned above are used in the majority of cases.

4. PAINTING AND DECORATING

Painting and decorating a structure is the last step in the construction process. The finish coating of the interior and exterior surfaces which require it, protects them from weather, air pollutants, dirt and dust, and at the same time can add to the esthetic effect of the structure. There are enough choices available in the selection of the proper coating to meet almost any condition.

4A. AVAILABLE COATING MATERIALS

There are two basic paints available for coating exterior woodwork. The one that has been used for a long time with completely satisfactory results is the oil base paint which contains a pigment of various proportions of titanium and zinc oxides and magnesium and calcium carbonate in a vehicle of linseed, safflower, and other oils. It is thinned with linseed oil and turpentine. This paint is used over a primer coat if the wood is new or can be used over an old paint coat which has firm adherence to the wood. It has long lasting qualities and excellent covering quality. The other exterior paint is acrylic latex which uses an acrylic resin emulsion as a vehicle for various metallic oxides and carbonates. It is thinned with water. It is usually recommended that acrylic latex be applied over an oil base exterior primer. It is advisable that the architect engineer and owner consult with one or more experienced painting contractors and carefully examine the surfaces to be coated before specifying this material for exterior woodwork. For wood shingles there are various types of creosote base stains available, as well as transparent stains, which protect the wood but allow the grain to show through. There are also many types of coatings available for exterior masonry walls. These vary from alkyd resins for floors, to vinyl latex for walls, or clear silicones which waterproof but do not color. Exterior metal requires protection. A rust inhibitive coating, oil based such as red lead or zinc chromate, should be applied over a clean surface before the final coating is applied. A two-coat oil based finish is recommended.

For interior surfaces the two most widely used finishes are the alkyd resins and the latex paints. The alkyd paints vary in composition. They consist of synthetic materials combined with pigments and vegetable oils such as soya, tung or linseed oil, and when dry, they form a resin which produces a hard surface. The latex paints use water as a thinner and vehicle. The exterior type uses a polyvinyl acetate base (PVA) with pigment added. The interior latex uses a polyvinyl chloride base (PVC). These paints are used for plaster, wallboard, or wood surfaces. They dry quickly and are washable when dry. They can be easily touched up.

For special purposes the architect engineer should be familiar with the use of coatings such as epoxy-polyester which forms a hard surface impervious to moisture and has good stain and soil resistance. There is a polyurethane varnish which can be applied to wood gutters or other surfaces that must be throughly protected against constant moisture. Polyurethane enamel forms a tough coating for heavy-duty concrete floors. There are other heavy-duty protective coatings available for conditions of high temperature and moisture.

4B. Workmanship Specifications

It is important that surfaces to be painted be first thoroughly cleaned of grit, plaster, dirt, dust, grease or other foreign matter. A ferrous metal surface that has been factory or shop coated should be washed with a cleaner such as benzine or gasoline before it is finish painted. Unpainted ferrous metal should be wire brushed, washed with benzine or gasoline, and given a coat of oil based rust inhibitor. Galvanized (zinc coated metal) must be cleaned and washed first as above and then washed again with phosphoric acid wash. This etches the zinc to allow for paint adherence. Concrete or masonry should not be painted until the last possible minute, when it is thoroughly dry. It must, of course, be carefully cleaned beforehand.

Plaster surfaces must be clean and any cracks or holes should be repaired with patching plaster. After the prime coat, any cracks that appear should be cut out and "spackled." (Spackle is a plaster-like material used for patching cracks).

Wood surfaces should be cleaned. All knots or sap spots should be covered with Knot Sealer (a Western Pine Association product) and then given two coats of shellac. Nail holes and other indentations should be puttied and sanded.

All surfaces must be absolutely dry and each coat should be dry before another coat is applied. Painting should not be done at temperatures under 50° or over 90°F. Gloss enamel or undercoats should be lightly sanded before the next coat is applied.

The decorative wall coverings such as vinyl wallpaper or fabrics must be placed on clean dry walls. These materials vary in quality and should be hung by skilled workmen in strict accordance with the manufacturer's directions. Careful attention should be given to the application of the recommended adhesive and the preparation of the wall surface. Such wall coverings are costly and heavy, and inferior workmanship could cause expensive repairs.

CONTENTS

1. Ceramic Tile—Its Qualities and Availability
   A. Its Use in the Home
   B. The Business Building
   C. Special Uses

2. Metal Toilet Enclosures

3. Accessories

17

CERAMIC TILE, METAL TOILET ENCLOSURES, AND ACCESSORIES

1. CERAMIC TILE—ITS QUALITIES AND AVAILABILITY

Ceramic tile is the most widely used wall and floor covering where resistance to fire, moisture, heat, weather, or soiling are vital. Ceramic tile lasts almost indefinitely and retains its color and texture under the most adverse conditions. (The ceramic mosaic tile floors that were laid in Roman times and buried under volcanic ash are still in excellent condition.) The architect engineer can use ceramic tile decoratively and it is often used for this purpose only. He also uses it where it must undergo hard wear. Ceramic tile is obtainable in an almost infinite variety of sizes, shapes, textures, and colors. Its hard smooth surface is simple to clean and it is resistant to abrasion.

1A. Its Use in the Home

In the private home or apartment building ceramic tile is most often seen in the bathroom and sometimes in the kitchen. As a bathroom wall tile the size and shape most frequently used is the cushion-edge, matt-glazed (dull) 4¼" × 4¼" tile which is laid over a plaster scratch coat on metal lath or waterproofed gypsum lath. The floor is most commonly a ceramic, unglazed mosaic (a pattern made of small pieces) tile that is laid over concrete which has been poured between wood joists (see Chapter 13), or over a concrete slab. In the more expensive home ceramic tile can be used as a kitchen floor, or

TILE, TOILET ENCLOSURES, ACCESSORIES

wall, or as a patio (indoor or outdoor) floor. In this case a quarry tile[1] is most frequently used. Because of its durability and complete resistance to moisture and insects, ceramic tile is used extensively for floors and wall covering in hot moist climates.

1B. The Business Building

Ceramic tile is used almost exclusively in business buildings for sanitary facilities and very often wherever floor or wall surfaces are subject to hard wear and soiling. The specification for the material calls for all tile to meet standards set by the National Bureau of Standards and Federal Specification for Tile. The wall tile can be cushion-edged or square-edged and the size available can vary from 4¼" × 4¼" to 6" or 8" × 6" or 8". The tile is usually specified as matt-glazed. Most sanitary facilities in business buildings are tiled up to the ceiling and have a ceramic cove base that finishes flush with the floor. The floor tile is specified as unglazed, ceramic, mosaic, machine made, paper mounted tile. Wall tile can be obtained in many sizes and in an almost infinite combination of patterns and colors. Quarry tile floors are often used in building main lobbies, reception areas, and other decorative spaces. Quarry tile is also used in kitchen and kitchen service areas. Although this tile is basically colored the soft orange red of terra-cotta, it can be obtained in other subdued earth colors of grays, and umbers, and tans. In such colors it is highly decorative. Quarry tile is specified as an unglazed, vitreous[2] body. It comes in various sizes and shapes as well as color. The workmanship specification starts with the preparation of the surfaces to receive the tile. For walls the gypsum plaster scratch coat, which is the usual base for tile, should be thoroughly wetted and then covered with a float coat or setting bed consisting of sand, Portland cement, and hydrated lime. This setting bed is carefully rodded or floated to an even surface flush with temporary screeds which have been fastened to the scratch coat with blobs of plaster or mortar. The tile is applied to this setting bed just before it attains its initial set. The tile has been soaked in water for at least a half hour and its back is covered with a thin coat of neat Portland cement before it is pressed firmly into the setting bed. The setting bed should be cut through to the scratch coat horizontally and vertically every four courses to prevent shrinkage. Avoid a larger setting bed than the corresponding area of covering tile which can be laid in the setting time of the bed, lest it set prematurely. When the tiles are thoroughly bonded they should be soaked and a thin mortar grout should be forced into the joints. The tile is cleaned after the grout has partially set.

[1]Quarry tile is made of thoroughly mixed shales and fire clays. The tile is extruded (forced through a die) before it is fired (baked) at high temperatures. It is usually terra-cotta colored and is wear-resistant and simple to clean.
[2]Vitreous in this case means "hard as glass," an effect produced by the fusing at high temperatures of the shales and fire clays.

TILE, TOILET ENCLOSURES, ACCESSORIES

For floor tile the underlying concrete bed should be carefully cleaned and thoroughly moistened before the setting bed is applied; this bed consists of lime, Portland cement, and sand and it should be carefully leveled to screeds. The mosaic floor tile which is usually pasted on paper sheets in the proper pattern is given a coat of Portland cement and pounded firmly into the setting bed just before it attains its initial set. After the tiles are bonded, the paper back is washed off and a thin nonshrinking grout is forced into the joints. Quarry tile is laid in the same manner as other floor tile except that in food service areas the pointing grout that is finally applied to seal the joints must consist of a material able to resist acids, alkalis, detergents, and grease.

When ceramic tile is of standard quality and is carefully installed it presents an almost impervious surface to moisture, heat, stain, and wear, and will last the lifetime of the building.

1C. Special Uses

Ceramic tile in its various forms is used by the architect engineer for outdoor plazas, decorative pools, and for such hard-use areas as school and factory corridors and stairways. For outdoor plazas it can be specified as a smooth faced quarry tile or a glazed ceramic matt-glazed tile with a glazed sanded surface. Ceramic tile can be obtained in patterned surfaces and in many shapes limited only by the taste of the architect engineer, owner, and by the budget. It is laid in a mortar bed over a reinforced concrete slab. The usefulness of the hard surfaced scratch and soil-resistant ceramic tile for school or factory corridors and stairways is self-evident.

2. METAL TOILET ENCLOSURES

The metal toilet compartment is used in all business and public buildings. The steel for the partitions is specified to be cold rolled, annealed, and patent leveled furniture steel. The steel should be 20 gauge for the partition and 22 gauge for the doors. The finish should consist of two coats of rust inhibitive primer and two coats of baked synthetic enamel. The best way to install such compartments is to hang them from structural steel in the ceiling. The cleaning and sanitation problem is minimal in this installation as compared to the compartments which set on the floor and serve as possible pockets for dirt or germs. The ceiling hung partitions are more expensive.

All the hardware for these compartments should be corrosion-proof, either stainless steel or chrome plated brass. The compartment walls and doors should be sound-deadened with a mineral or glass fiber cellular material.

TILE, TOILET ENCLOSURES, ACCESSORIES

The doors should be equipped with latches and gravity hinges. These hinges are so placed that the door cannot remain open. A roller under spring tension is attached to the door. It travels on an inclined cam and closes the door gently and positively. All movable portions of the closing hardware are on roller bearings.

Toilet compartments vary in width from 32 to 36 inches and are normally 56 inches deep. The distance of the ceiling-hung partition above the floor is about 12 inches and the partition and door are about 58 inches high.

3. ACCESSORIES

The bathroom or toilet room accessories vary in their material, depending on their use. In the home they very often consist of ceramic material set into the ceramic tile. In a public building they are stainless steel or chrome plated brass. The usual accessories consist of the following: paper towel dispenser and waste receptacle combinations: toilet paper holders; purse shelves or hooks; continuous shelving over lavatories in men's toilets; individual metal framed mirrors and shelf combinations (mirrors in ladies rooms should not be placed over lavatories if possible); soap holders; grab bars; clothes hooks, and sanitary napkin dispensers, and disposal units. The architect engineer has a wide choice of designs. He should specify by material standards of ASTM or Federal Specification. The finish should be specified as to thickness and brightness. For instance, chrome plating should be applied over a coating of nickel over a brass base. Over a steel base the chrome should be applied over a coating of nickel over a coating of copper. The accessories should be simply designed so as to be easy to keep clean and must be carefully and sturdily fastened to withstand a great deal of abuse. Some mounting heights for accessories are: toilet paper holders—30 inches to center; robe hooks—66 inches; towel bars—44 inches; soap holders—60 inches.

It should be emphasized that everything mentioned in this chapter, whether tile, enclosure or toilet accessory, is subject to hard wear, excessive moisture, rough handling and abuse, and must be of good material, and carefully installed to withstand this usage.

CONTENTS

1. THE PURPOSE OF ORNAMENTAL METAL
   A. Where it is Used
   B. The Material Used
   C. The Fabrication and Erection

2. HOLLOW METAL DOORS AND FRAMES
   A. The Material Used and the Fabrication and Erection

18

ORNAMENTAL AND HOLLOW METAL WORK

1. THE PURPOSE OF ORNAMENTAL METAL

The ornamental metal work in a building consists of all the metal work that is used not only to perform certain structural purposes but also for architectural effect. Unlike miscellaneous metal (Chapter 15), which is part of the working structure but is hidden in work spaces, ornamental metal is meant to be seen by the occupants and the general public. The architect engineer therefore pays special attention to ornamental metal design and construction and makes sure that it helps to carry out his overall design.

1A. Where it is Used

The private home or multiple dwelling uses very little ornamental metal. It may be used as a wrought iron railing or as suppport for a portico roof at a front entrance. It is also used as a stair rail or other metal ornamentation in certain types of architecture. The architect engineer should be sure to carefully specify his requirements. In the office building, some of the uses for ornamental metal would be as follows: the metal enframement of the ground floor lobby, complete with all suspension steel and steel subframing; the finished metal framing members including glass and glazing and all interior and exterior finished trim including air supply enclosures; all revolving doors, tempered glass doors, and entrance doors complete with glass and glazing;

ORNAMENTAL AND HOLLOW METAL WORK

hollow metal doors and their frames that are of such metals as stainless steel, bronze, or aluminum; all ornamental partitions; all stainless steel or nonferrous metal door saddles, sills, nosings, or corner guards; all stainless steel or nonferrous metal column cladding; all ornamental ceiling, and floor trim, and other horizontal and vertical trim in connection with interior marble, mosaic, or other wall facing; all stainless steel or nonferrous metal railings; the furnishing and installation of all hardware to be used with any of these items of work and the caulking and sealing between the installation and the surrounding construction. Ornamental metal work may include porcelain enameled steel or aluminum and anodized or duranodic finished aluminum.

1B. THE MATERIAL USED

The materials used are specified by standards set by ASTM or others and by gauges of metal. If aluminum is used, the proper alloys for the various structural members are specified. The porcelain enameled surfaces are specified by degree of gloss and an acid resisting rating as set by the Porcelain Enamel Institute. The surface finishes for stainless steel, aluminum, or other metal are specified. The color and thickness of the anodic or duranodic finish is specified. The nonferrous door saddles are usually of the cast abrasive type. The exterior saddles, which must be corrosion resistant, should contain 20% of leaded nickel silver combined with the aluminum, and the interior saddles should be of special hard alloy aluminum. All exposed surfaces should contain not less than 2 ounces per square foot of evenly distributed aluminum oxide abrasive granules embedded at least 1/16 inch into the surface.

1C. THE FABRICATION AND ERECTION

The general directions for the workmanship in fabrication and erection call for such things as carefully mitred and fitted joints; concealed screw heads; preassembly at the shop to make sure that the parts will fit together in the field (this is very important—the fabricating shop may be a thousand miles from the job and field correction of a bad assembly is very expensive, and slow, and cannot compare in quality with shop work); the proper protection of the metal surfaces during shipping and erection; a requirement for flatness of the members; a requirement that the welding conform to standards set by The American Welding Society; and directions for caulking. The setting of glass is important in any ornamental metal frame and the directions for the workmanship and materials to be used include the neoprene gaskets, the interior nonstaining butyl rubber glazing compound, and the exterior thiokol base pressure gun applied sealant. The revolving doors should be of the panic-proof braceless type with the wings folding on each other in the

ORNAMENTAL AND HOLLOW METAL WORK

lines of egress with an adjustable collapsing pressure of from 60 to 175 pounds. This means that in the event of a panic the wings of a revolving door can be collapsed under pressure to allow escape to the outdoors. The revolving mechanism of these doors is controlled by a speed regulator which prevents spinning at uncontrolled speed. The glass is of the safety type which consists of two sheets of polished plate glass bonded by a middle layer of plasticized polyvinyl butyral resin. There are also directions for the erection of store front metal (the metal enframement of store fronts is a part of ornamental work), and its glazing, and the fabrication and erection of stainless steel panels, column cladding, or other uses in kitchen or laboratory areas.

2. HOLLOW METAL DOORS AND FRAMES

Hollow metal work is mainly concerned with doors and door frames located in fire-resistant buildings. Building codes require that openings in fire-retarding walls or partitions be only slightly less fire-retardant than the wall. For instance, a 3-hour door would be required in a 4-hour wall. All doors to fire exits, legal stairways enclosed by fire walls, elevator shaft openings or openings between fire divisions[1] in a building must be approved labeled doors. The label is granted by the National Board of Fire Underwriters for doors constructed in accordance with its specifications. The label gives the fire rating of the door in hours.

2A. THE MATERIAL USED AND THE FABRICATION AND ERECTION

Hollow metal doors and frames, are made of furniture steel. In ornamental work the hollow metal may be of stainless steel or a nonferrous metal. This name is applied to cold rolled, annealed, patent leveled, and bonderized sheet steel that is free of scale, waves, or other defects. The interior door most often used has a 1½-hour rating and is constructed of two outer sheets of 18 gauge furniture steel continuously reinforced with 20 gauge interlocking members on 6-inch centers. All construction joints are welded and ground flush. The hollow between the outer sheets is filled with a sound-deadening and heat-retardant filler. When double doors are required they must be rabbeted or have an astragal (an overlap between the doors).

Hollow metal frames are made of furniture steel which must be 14 gauge in order to be labeled. All the frames must have at least three steel corrugated anchors on each side that are welded to the frame and which must

---

[1] A fire division is a fire retardant wall built between sections of a structure to prevent the rapid spread of fire. For instance, the wall between a garage and house or the wall around a fire stair is a fire division.

ORNAMENTAL AND HOLLOW METAL WORK

be securely built into the surrounding wall. The frames are made with welded joints and the architect engineer should specify that they be reinforced at the head and that the hollow sides be slushed in solidly with cement mortar when they are built into a wall.

All doors with a fire rating must be provided with automatic closers and must be reinforced for the application of these closers as well as for the other hardware such as locks, door knobs, and hinges. The inside surfaces of all doors and frames usually receive a shop coat of baked-on rust inhibitive primer and the outside surfaces receive a coat of baked-on enamel filler and a baked-on prime coat.

A code requirement for all such doors is that they cannot be locked on the side from which egress is to be made, but can be locked to prevent access from an approved exit stair to a corridor or rented space.

19

FINISHING HARDWARE

The specification for such trades as hollow metal, ornamental metal, carpentry, curtain wall, miscellaneous metal, and many others, always states that the material to be supplied and installed includes hardware. In these cases hardware includes brackets, bolts, screws and hinges, fastening, locking and holding devices. In addition to this miscellaneous hardware, which is a part of many trades, there is also a category in construction called *finishing hardware*.

Finishing Hardware. Finishing hardware is what most people think of when the word hardware is mentioned. The finishing hardware is what everyone sees and it consists of such items as door closers, door stops, kickplates, push plates, push bars, panic bolts, lock sets, latch sets, door openers and thresholds. The architect engineer should be familiar with the materials and patterns that are available and choose hardware that is consistent with his design concept because the public and the tenants are always aware of this construction material.

The materials chosen must be of corrosion-resistant material such as stainless steel, bronze, aluminum or chrome plating. Items such as door handles or knobs, push bars or push plates are handled constantly and must retain their appearance not only through this handling but also through constant cleaning and polishing. If chrome plating is used it should be over nickel plating over brass. Even corrosion-resistant materials should have finished surfaces that will be resistant to abrasion, denting, or other mishandling. Many specifications call for the employment of a member of the American Society

FINISHING HARDWARE

of Hardware Consultants by the finishing hardware contractor, in order to advise the architect engineer and owner on the proper selection.

The finishing hardware is usually specified on a schedule which gives acceptable manufacturers for each category and states certain general requirements for quality of material and installation. The schedule then mentions each door opening by a class number and states precisely what hardware is to be furnished for all doors of that classification. The general requirements would be as follows—for each approved manufacturer the specification mentions an approved lock or latch set by catalog number. The material of the door knobs, lock face plates, strike plates and cylinder faces is specified as a non-corrosive metal. The cylinders in a large building that must be master keyed and grand master keyed are specified to be provided with six pin-tumblers. The hinges for the interior are usually cold rolled wrought steel and for the exterior are a noncorrosive metal with nonremovable pins. Because the hinges have to bear the weight of the door, hinge sizes and the number of them to be used are specifically mentioned by door thickness and size; for instance, a door 1¾ inches thick and not more than 3 feet wide by not more than 7'6" high should use three 4½" × 4½" hinges.

The keying system for any large building or building complex must be carefully studied and must provide the maximum ease of access with maximum security. A typical manner in which this is done is as follows:

1. A grand master key setting that will open all the locks in the building. Such keys are very carefully accounted for. Provisions should be made to have one such grand master key available at all times.
2. A master or watchman setting that will open entrance doors and certain specified locks. Such a key is necessary to enable watchmen to enter and check any sensitive areas.
3. A mechanical master setting to open all mechanical spaces. This key is given to the engineer's crew, to electricians, and other mechanics. The watchman's master key can also open any or all of these spaces as designated.
4. An individual key setting for each lock.
5. A stair control key for cylinders of all stair doors.
6. All doors to the same space are keyed alike.
7. All exterior doors are keyed alike.
8. All separate office doors are (of course) keyed differently. It is wise to have the architect engineer help the owner set up a key control system which keeps all keys inaccessible to anyone but trusted personnel, and at the same time allows the quick replacement of a lost key, or a change in a cylinder setting if necessary.

The architect engineer must be familiar with the use of special-purpose finishing hardware such as automatic door openers and closers. These automatic devices can open single or double doors by swinging them open or sliding them open. They can be actuated by an electric eye located in the floor and this device is often used when hand or power trucks have to move from one part of a plant or building to another through normally closed doors. These automatic openers may also be actuated by stepping on a pad in front of the door, or by pushing a plate or bar set on the door. The mechanism that opens and closes the doors may be electric, pneumatic, or hydraulic. The actuating mechanism starts a hydraulic or air pump, or an electric motor which moves the door open and shuts it again or opens it only and lets a spring do the closing. Certain types of entrance doors use floor hinges that are set into a concrete base under the door saddle. These floor hinges contain heavy springs which close the door and are controlled by a compressed air check so that the door closes slowly. All stair and many other exit doors are specified as fireproof self-closing (FPSC). These doors are closed by the use of an automatic door check. These checks close the door by means of compressed hydraulic fluid, compressed air, or by heavy torsion springs immersed in oil which move against an air or oil check valve.

There are many styles of door knobs, lever handles, and lock faces which may even be silver or gold plated if desired. There are unit locks in which the key is inserted into the knob of the door.

The finishing hardware industry knows that its product is on view at all times and has accepted the challenge by making available a variety of designs and metals.

CONTENTS

1. THE REQUIREMENTS FOR A BUSINESS BUILDING
    A. The Modular Plan
    B. Required Flexibility and Bidding Procedure
    C. Interior Partitions
        1. The available choices
            a. The metal partition—design and quality requirements
            b. The dry wall partition

2. OFFICE PLANNING
    A. Setting up Standards for Space and Facilities

20

FINISHING THE INTERIOR FOR OCCUPANCY

1. THE REQUIREMENTS FOR A BUSINESS BUILDING

This chapter will examine the architect engineer's role in the completion of a business building for tenant occupancy. In a private residence or apartment building the plan from which the structure is built defines the various rooms and the separate apartments, and except for the finished wallpaper, paint color, and color choices of appliances, the structure is carried through to completion when it is ready for occupancy. The factory building should be a shell enclosing the assembly or flow line that produces the manufactured article. The special-purpose medical building or laboratory building is also usually built as a completed product. The office building, however, is built as a shell with open floors and these floors must be made ready for occupancy *after* the building is completed. For a large tenant or a single occupant it may be possible to build a great many of the tenant's requirements into the building as the general construction progresses. Even in such cases, however, there are constant changes in layout as the building progresses, and it has sometimes been found necessary where a large tenant or single occupant is concerned to forbid any changes in plan at least three months before the scheduled completion, if the tenant is to ever move three months before the scheduled completion, if the tenant is to ever move in.

FINISHING THE INTERIOR

1A. THE MODULAR PLAN

The conduct of business requires constant changes in procedure and the consequent reorganization of space is taken into account by the architect engineer. The design of a business building should be based on a modular plan which assures the owner that space can be divided into multiples of this module. Lighting fixtures, underfloor electrical ducts, window spacing, and air conditioning units should be part of the system. This will enable a tenant to subdivide his space so that every office can have all available facilities without major alterations. The architect engineer very often is retained by the tenant or owner to do the planning for the tenant. If he does not do the architectural layout work he must do the mechanical planning and he should insist on doing this to preserve the mechanical integrity of the building as a whole.

1B. REQUIRED FLEXIBILITY AND BIDDING PROCEDURE

However, designing the building to a module is just the beginning. The ceiling system and the lighting fixtures must be designed so that they are completely flexible and so that a partition can be erected on any modular line. The architect engineer should also, in conjunction with the owner, arrange to have unit prices included in the contractors' bids on all the facilities that must be installed in tenant space. Various areas of the country have different standards for the amount of work that owners furnish to tenants. Having these "building standards" in mind, it is wise for the bid document to call for the contractor to furnish a certain amount of such facilities as part of his original bid. For instance, if a local standard calls for 1 linear foot of partition for every 15 square feet of space, the bid request may be for 10,000 feet of partition to be included in the bid on a 150,000-square-foot building. The same kind of proportion, depending on the owner's wishes, should be observed for other items such as electrical outlets, thermostats, perimeter air conditioning units, ceiling diffusers, and lighting fixtures. The request for bid should enumerate all of these items, and others where necessary, and ask that a certain number be included in the base bid. One of the most important parts of a construction contract concerns its flexibility, and the architect engineer and the owner must keep their options open. The contract agreement should allow the latest decisions possible as to where lighting fixtures go, the location of perimeter air conditioning units, interior mixing boxes, and the location of thermostats and ceiling diffusers, among other things.

1C. INTERIOR PARTITIONS

The interior partition in a business building is a very important part of the appeal of the building to an owner occupant or a tenant. In present-day business, changes in office procedures, new products, conglomeration and many other influences may cause constant rearrangement of an entire office complex. The partitions that separate the various working spaces must be able to be changed many times and reused over and over again without suffering damage. This constant changing can be illustrated by one building occupied by a single tenant in which about 20% of the partition is moved every year. In this particular building this involves 12,000 linear feet. Sometimes the same partition is moved several times in a year, if this move expedites the way of doing business. The partition must therefore be quite sturdy and simple to move. It must in addition be good looking, sound proof, and easy to keep clean.

1C1. *The Available Choices*

The architect engineer has a choice of many kinds of partitions. Before choosing the kind to use he should consult with the owner as to the use the building will be put to; whether there will be many small tenants or several large ones, or whether the building will be occupied almost exclusively by one tenant. It is strongly recommended that room-sized sample mockups be erected and studied before a decision is made. The cost of partitions varies widely and should be discussed. The two most widely used types of moveable partitions are metal (steel or aluminum) and the dry wall type which consists of various combinations of wallboard made of gypsum or compressed fibers.

1C1a. *The metal partition—design and quality requirements*

The metal moveable partitions consist of panels made of galvanized, bonderized, cold rolled steel laminated to a center core made of honeycombed fiber or glass fiber for better acoustic quality, or solid compressed hardboard for heavy impact resistance. The panels come in standard modular widths of 24 to 60 inches with 6-inch intervals between. Heights vary from 7'2" to 9'2". Doors can be obtained in various standard widths. These laminated panels are attached to posts made of steel or aluminum and are held at the floor and ceiling by continuous metal runners. The methods of fastening the panels to the posts, and to the runners, and to the door frames vary. A careful study should be made to ascertain that the fastening is sturdy and can hold the partition rigidly against reasonable impact. The fastening must also be simple so that it does not require highly skilled labor to take down and erect. The fastening should require no more tools than a hammer, screwdriver, and wrench.

Metal partitions come in thicknesses from 2¼ to 3 inches. There are partitions with concealed posts available; they can be obtained with solid panels, half glass, all glass, or a solid panel with a borrowed light above. This last variation is significant. It provides visual privacy but allows light spill between spaces. Partitions can be obtained with recessed heads, bases, posts, or with feature strips at all of these places. There are partitions available with exposed aluminum posts, bases, and door frames. The panels can be factory-finished with vinyl or plastic laminates, wood veneer, or baked-on enamel. Railings and low partitions are also available in metal in various modular widths. These are usually 3′6″ high and with glazing above they run to 5′6″. This height provides visual privacy.

A typical specification for a good quality metal partition (one that has been moved at the rate of 12,000 feet per year by semiskilled labor and is still good), would contain the following design requirements:

Soundproofing. Each component of the moveable partition system shall be constructed so as to provide a transmission loss of not less than 32 decibels in the frequency range between 150 and 4800 cps. (See Chapter 14 on "Sound Conditioning.")

Flexibility. The partition system must provide complete flexibility, allowing the removal or installation of any door, glass, or panel unit without disturbing adjoining units in any way.

Appearance. The installed partition must be free of any exposed fastenings.

Adjustments. Any deviation in floor or ceiling height must be taken up in the partition base and head section. Provision must also be made for horizontal adjustment.

In this particular case the partition framing is aluminum, which is often chosen because of its light weight. The aluminum posts and trim are of extruded aluminum sections. The hollow metal doors are of 18 gauge cold rolled annealed sheet steel free of any waves. The metal panels are stiffened and reinforced horizontally and vertically and the core is filled with sound-deadening, heat-retarding material. The erection specification, after calling for the usual straight, plumb, and rigid erection, also calls for a sound test after completion. This is not often done and there are cases where unsuspected sound leaks between offices have created serious trouble. In any case the architect engineer must first specify a sound-deadening blanket or rigid board to be installed between the top of every "privacy" partition and the underside of the structural floor above. In addition it is recommended that a sound generating device be run in one office and that any openings through, over, under, or around the partition be caulked until the sound transmission loss reaches the desired level. This, of course, need only be done in particular areas but it is important that it is done where necessary.

The quality of the partition is determined by certain standard performance tests some of which are as follows. Baked enamel finishes are tested for stain resistance to commercial cleaners. They must withstand several commercial cleaning solutions that are left on the finish for 24 hours, with no change in color or hardness. They must be able to be washed clean after exposure to 4 hours of tincture of merthiolate, or 4 hours of hair oil, or black ink, or even 1 hour of synthetic perspiration. The finish must withstand tests for hardness or impact with no cracking. It must withstand 400 hours of accelerated weathering or 400,000 brush strokes while immersed in a 5% solution of tri-sodium phosphate (a commercial cleaner).

A metal partition that can meet all these tests and specifications will last indefinitely with very little maintenance and can be installed and reinstalled inexpensively.

1C1b. *The dry wall partition*

The dry wall type of moveable partition is used extensively. It is not as expensive as the metal moveable partition nor as sturdy or capable of being moved as often. The initial cost, however, can be very important to an owner especially where he must furnish a great deal of partitioning to a tenant without charge as a "building standard item." These partitions can be good looking, sturdy, and of variable acoustical quality. They can either be moved as individual panels or in sections. Because of their lower initial cost, dry wall partitions are being used more often in a greater number of high-quality office buildings.

Most dry wall partitions are gypsum board that is combined in various thicknesses to produce required acoustical qualities, or strength, or fire rating. Like the metal partition, it is mounted on floor and ceiling runners. There is a type of dry wall partition that comes in panels, and is fastened by hidden clips to concealed metal studs. A typical specification for such a dry wall fastened to metal studs would contain the following provisions: Furnish and install metal stud and gypsum wallboard partitions complete with partition end fillers, sound-absorbent barriers above hung ceiling (where noted), and within window induction units where partitions abut them. The gypsum wall board is to be 4' × 8' and of various thicknesses for acoustical as well as for other purposes. The stud is to be galvanized steel channels 2 inches wide. The metal floor and ceiling runners are to be galvanized 16 gauge steel. Hollow core is to be filled with sound barrier of batt-type glass fiber or equal. The floor and ceiling runners are to be securely fastened at intervals of not more than 24 inches and the studs are to be 16 inches on center and at every door frame. The wallboard should be applied to the studs so that the joints on the opposite sides occur at different studs. The joints are to be taped and covered with joint compound brought to a feathered edge or the joints between the boards can be covered by decorative battens.

Some of the dry wall partitions do not have metal stud supports and their structural strength is due to the fact that they consist of several thicknesses of board that are glued together. To get an idea of this strength, a triple solid wall consisting of three ¾-inch-thick gypsum boards with staggered joints or tongue and groove joints will handle fairly heavy impact loading and normal wear in unsupported lengths up to 12 feet and an unsupported height up to 14 feet. There are dry wall partitions that can serve as demising walls (walls separating different enterprises). They are soundproof and have a 2-hour fire rating which is acceptable in most codes. Such walls may consist of two layers of 1-inch-thick wallboard covered on both sides by a ½-inch-thick layer of wallboard with all four thicknesses totaling 3 inches. Another version of a demising wall consists of two outer layers of ½-inch board each glued to a ½-inch coreboard with the coreboards separated by a 1-inch air space. The entire thickness is also 3 inches. For complete acoustical protection there is a demising wall consisting of two layers each 1⅞ inches thick. Each layer consists of ⅝-inch-thick coreboard strips glued to a layer of ⅝-inch board on either side with a 1½-inch air space between, the entire thickness being 5¼ inches. The usual office partition wall is 2¼ inches thick and consists of a ⅝-inch board on either side glued to a series of 1-inch coreboard strips. The surface can be prefinished in the factory with vinyl, wood veneer, or plastic laminate.

Some partitions can be obtained with recessed cap and base molding, and offer a wide choice of door frames and doors. Dry wall construction is also used for rails or low partitions. It can be obtained in various widths of tongue and groove panels with V joints, or it can be butted solidly with taped joints.

2. OFFICE PLANNING

The planning of the interior working arrangement starts with the programming of the tenant's requirements. The first step in such programming is a head and equipment count. The tenant and the architect engineer must agree on a space standard by which every employee is allotted a certain amount of space in which to perform his function.

2A. Setting up Standards for Space and Facilities

A study of many company space standards provides an average as follows: (The figures vary somewhat depending on the building module.) The head of the firm and his senior officers are allotted whatever space they require to present the proper image of the company.

A vice president and department head	From 295 to 375 square feet plus 100 square feet of access space
Lower echelon officers	From 215 to 280 square feet plus 100 square feet of access space

Managerial employees	From 140 to 210 square feet plus 50 square feet of access
Desk positions for clerical employees	42 square feet or 85 square feet depending on whether they require a chair for a visitor
Conference rooms	20 square feet *per person*
Office equipment	Width times six feet

These space standards are multiplied by the people in the various categories and this area plus area for equipment, reception rooms, conference rooms, etc., gives the total space requirement.

The next step in the office planning procedure is the obtaining of the work load of the different departments or sections and the determination of the paper flow between the sections. Once this is established the planner can start to lay out desk spaces and office spaces to promote an orderly flow of work. The space layout follows the modular plan so that office partitions, lighting fixtures, and air conditioning facilities are available for every office and so that every desk space has enough light, and air, and a convenient electric outlet. The architect engineer and owner should now have at their disposal all the tenant facilities that were requested and are part of the original construction bid. The office layout that stays within these limits is pleasing to the owner and to the tenant who very often has to pay if he exceeds his allowance for partitions, outlets, switches, or other facilities. The planner must also make sure that soundproof partitions are placed where they belong; that a thermostat is placed in the proper offices and that either the perimeter air conditioning units or ceiling mixing boxes are installed so that they provide the proper supply for every space.

There is another type of office layout, if it can be so named, that places people in work clusters throughout an open floor with the only separation being low-railing type partitions, or planter boxes, or low screens of various types. The clusters are usually round in shape and the desks within them do not form any geometric pattern but are seemingly placed at random. To avoid the overhearing of conversation from one cluster to another or even within a cluster the designers have supplied background noise of sufficient volume to prevent this but not to prevent normal conversation. Several layouts of this kind are in existence and look attractive. They may work well for some kinds of operations.

The good office design with its potential to make people happy and efficient is the final step in the planning of a successful business building.

21

LANDSCAPING

The proper landscaping of a building site in order to develop a harmonious setting for a private home, apartment building, office building, or any other structure is a very important part of the architectural concept. The architect engineer should be aware of what is involved in landscape planning. Chapter 1 on "Site Planning" discusses the general principles involved in locating a structure on a piece of land. The chapter discusses general principles which the architect engineer must keep in mind in the orientation of the structure, investigation of the soil, and the general knowledge and awareness of the entire area where he is planning to build.

The landscaping of a site involves, of course, much more than this. The relationship of the planting, walks, and utilities, to the structure, and the screening of a structure from adverse influences should be a part of the architect engineer's plan. The grading of a building lot to present or screen a particular view, or the planning of an entire subdivision requires a knowledge and a sympathetic understanding of human requirements. We have all seen subdivisions built on flat former farmland where they simply square off the streets, pave them, build row on row of little houses cheek by jowl, and plant one tree in front of each house. (Some developers don't even plant one tree.) It would cost no more to curve the streets, leave an occasional tree-planted green plaza at an intersection, and plant trees. The developer might have to add to the sales price of each house but he would also be satisfying a human need and be adding some charm which would make friends and influence people. We have also seen what happens when developers

LANDSCAPING

strip hillsides of shrubbery and cut wide swathes through woods and natural rises in the land. Nature gets her revenge in creating mud torrents after heavy rains that wash entire houses away. The deadly monotony of long vistas of concrete or black top roads certainly does not add to the feeling of openness or identification. Such ruthless tearing of the earth has been called "rape by the carry all."[1] Many communities have taken cognizance of this and are prohibiting these asphalt jungles. The planner can take advantage of natural slopes and stands of trees to curve the roads. He places houses in relation to sun direction, prevailing winds, and the slope of the land. A survey of the site should include a study of the zoning and restrictions on height, side yards and set backs, surrounding roads and their traffic pattern; the high and low spots with evidence of swampy areas or ledge rock; the sun and wind directions; existing trees; the direction and distance of any nuisance areas; the views from various locations. With the use of this and other information, the planners can choose logical building sites. They can make some roads dead ends with a turn around for fire protection and snow plows. This will provide quiet streets where children and pets may be reasonably safe. The house on such a street and on such a planned site will be quickly bought.

In an effort to help such planning many communities have created special zoning enclaves where land areas permit. These encourage the architect engineer and owner to get away from the gridiron street pattern. This specially zoned area can be used for multiple housing or small office buildings. Curving the roads conceals parking areas and the proper treatment of trees and shrubbery enables such enclaves to be placed near or in residential zones. These attractive sections add rather than detract from the area.

Buildings in large cities, where land is expensive and directions are fixed by the street pattern, present a challenge to the architect engineer and landscape planner. The Zoning Code has been changed in most communities and will give bonuses in allowable rentable area in return for using less land. The land that remains uncovered by the building can be used to enhance the entire architectural concept. Many studies should be made of the direction from which the public will approach the site, the scale of the area compared to the building and the surrounding structures, and even the type of people who pass by and use the open area. (Are they office workers, local residents? Or is the building in an area in which undesirable characters are likely to congregate?) Much can be done with a small area by using planting that will stand up to city fumes and heat; by using planting material that is frankly scaled down, instead of high trees which simply cannot compete with their surroundings; and by the use of water in fountains or reflecting pools.

[1] A carry all is a very large bulldozer-scraper that can scrape and carry away dozens of tons of earth at one load.

LANDSCAPING

Channel Gardens, Rockefeller Center, N.Y.C. Courtesy Rockefeller Center, Inc.

The effective combination of water with planting or simply as an architectural concept goes back many centuries and is still valid. The fountains of Rome are known throughout the world. There are many excellent examples of the use of water in architecture all around us. The channel gardens of Rockefeller Center in New York City are a skillful combination of planting and running water. Here series of pools lead down a gentle slope, each fed by a jet of water and surrounded by frequently changed planting. The open plaza in front of the Equitable Life Building in Chicago has a fountain in a circular pool, skillfully arranged planting beds, and seating space, all of which provides an oasis in the middle of a large city. The paving of the plaza becomes a part of the landscaping by the use of warm colored brown brick. Many architect engineers use sculpture to enhance their open areas. A striking example is the sculpture garden of the Museum of Modern

LANDSCAPING

Prometheus Fountain. Rockefeller Center, N.Y.C. Courtesy Rockefeller Center, Inc.

Art in New York City. Its running water and fountains and beautiful weeping beech trees, dominated by superb sculpture, give a sense of serenity and rest in the midst of a roaring city. Some architect engineers have made use of a sloping street and set their building on a podium, reached by low wide stairways interspersed with fountains, plantings, and light. In this connection the skillful use of lighting can make the open area a pleasant place during the night hours. All of these landscaping possibilities make excellent public relations.

The Ford Foundation Building in New York City has its entire landscaping concept within the building. In an area approximately 120′ × 100′ it encompasses terraced gardens, running water, and trees and shrubs. It is a unique and beautiful example of imaginative architecture as expressed in landscape planning.

The architect engineer and landscape planner need only look around them with a thoughtful eye to provide man, for whom they are building, with beauty.

CONTENTS

1. GENERAL USES

2. THE PRIVATE RESIDENCE
   A. Windows
   B. Doors
   C. Trim, Paneling, and Cabinets

3. THE OFFICE BUILDING—THE USE OF MILLWORK
   A. The Required Quality of Material
      1. Doors
   B. Required Workmanship

22

CABINET WORK AND MILLWORK

1. GENERAL USES

Almost every structure requires some cabinet work and millwork to make it suitable for human needs. The single-family house requires such work as wood doors, door frames, windows, baseboards, stairs and stair rails, paneling, kitchen counters and cabinets, closets, and other conveniences. The apartment building requires closet shelving, kitchen cabinet and counter work, door trim and interior doors (where the fire code allows). In manufacturing plants the office quarters require cabinet work and millwork. In office buildings this work includes doors, paneling, shelving, and many other conveniences and decorative work to provide comfortable, functional, and attractive quarters.

2. THE PRIVATE RESIDENCE

In the private residence the architect engineer should become familiar with the available choices in windows, doors, moldings, cabinets, and other finished millwork. The lumber itself should conform to standards set by Federal Bureau of Standards or the National Hardwood Lumber Association. Some of the soft woods generally used are Douglas Fir, Ponderosa Pine, and Western Pine. All lumber is specified to be kiln dried with moisture content set by standard practice. The architect engineer should also be-

CABINET WORK AND MILLWORK

come familiar with the standards set by the Architectural Woodwork Institute. This organization publishes a manual as well as a Wood Characteristics Table which sets forth the use recommended for every kind of available decorative wood.

2A. WINDOWS

Windows for the private home can be obtained in almost every possible combination of size and opening. To mention some: there are awning sash and hopper sash (both open outward, one swinging from the top and the other from the bottom of the opening); there are casement windows, sliding windows, bow windows, and bay windows and, last, the popular double-hung windows. All windows should be specified to be treated with a toxic, water-repellent, wood preservative. They should be fitted with built-in weather stripping. Casement windows can be fitted with storm windows and screens. Many windows can be double glazed for insulating purposes. There are windows with built-in moveable metal louvers for protection against hurricanes or vandalism. Jalousie windows consist of tempered glass movable louvers. When they are closed there is glass to glass contact. They are recommended for porches and other quarters that do not require high insulation against severe weather. All windows serve a definite architectural function and should be selected accordingly. Windows also serve to provide light, ventilation, and protection from weather. They should be conveniently located and easily operable—many houses have a double-hung window over the kitchen sink where a housewife spends so much of her time—this window is difficult for her to open whereas the provision of a swinging casement that can be opened by a handle will show forethought and concern on the planner's part.

2B. DOORS

The door in a properly built residence is generally a panel door. The usual thickness is 1⅜ inches for interior doors and 1¾ inches for the exterior. The specification should call for kiln dried close grained pine such as ponderosa. A good panel door will have solid stiles and rails that are glued and doweled. In the normal 2'6" to 3'0" wide door the rails should be about five inches wide for the top and ten inches for the bottom. The panels should be at least 1 inch thick on inside doors and 1⅜ inches thick on the exterior. The exterior doors should be treated with a toxic, water-repellent preservative. There are many combinations of standard panels to choose from. Many residences use flush doors, both hollow and solid core but the greatest use for these is in office buildings. Their construction will be described later in this chapter. Closet doors can be sliding, folding, louver, or flush panel. The swing of doors should be carefully planned throughout.

2C. Trim, Paneling, and Cabinets

While the doors and windows are very important parts of the architectural whole there are many other cabinet and millwork items that are both decorative and necessary. Kitchen and bathroom cabinets and counters can be imaginatively and conveniently arranged and can be made handsome by using available stock parts. The choice of cornices, moldings, and door frames can add charm to the detail of a house. Wood paneling can now be obtained at reasonable prices and it can be decorative. It comes in ¼-inch-thick sheets that are 4 feet wide by 7, 8, or 10 feet long. The sheets consist of a layer of hardwood plywood to which a rotary sliced veneer is pressure-glued. For the veneer there is a choice of such woods as walnut, oak, elm, cherry pecan, mahogany, birch, and others. The veneer surface is prefinished and the panels can be installed over wood grounds. The surface can be scored in random widths or made to look like planking. This type of paneling is today's answer to high labor costs and the shortage of skilled workers. For the fine residence or office there are carved wood room dividers or screens. For special conditions pressure-treated wood is available which is resistant to fungus, decay, and termites. The architect engineer can also obtain wood columns and pilasters that are fluted, round, or oblong. The bases and capitols can be of any Greek order or no order. If the architect engineer has a client who specifies it, barnboards, beams, and shingles that are guaranteed to be at least 50 years old are obtainable. The woodworking industry aims to please!

3. THE OFFICE BUILDING—THE USE OF MILLWORK

Wood cabinet work and millwork is used extensively even in fireproof buildings. The work furnished under this trade might include: natural veneered and plastic laminate surfaced wood wall paneling; natural veneered and plastic laminate surfaced wood counters; corkboard and chalkboard faced wood paneling; rolling wood partitions; wood doors and folding doors; hollow metal doors that are designated to receive wood veneer surfacing; closets complete with hanging rods and hat shelves; wood shelving; wood nailers and grounds or the installation of this work; fireproofing of all work where required by code.

CABINET WORK AND MILLWORK

3A. THE REQUIRED QUALITY OF MATERIAL

The quality of the lumber is generally the same as previously mentioned in this chapter. The quality of the plywood, which is used as backing for veneer paneling or plastic laminate, is carefully defined. The specification calls for a solid staved and cross-banded core, the staves of nonresinous hardwood not over 1½ inches wide and of thickness as called for; and the cross bands of poplar, birch, or maple of random width and not less than $\frac{1}{16}$ inch thick. The veneer which is pressure-glued to the plywood backing can be rotary cut or sliced. Sliced veneer gives the architect engineer the opportunity to match color and grain. It is more expensive than the rotary cut. The treatment of all wood is specified both as to preservative and fireproofing. All wood that is to be in contact with masonry, concrete, or structural steel should be pressure-treated with a solution such as pentachloro-phenol. All wood requiring a fire rating must be completely impregnated with a noncombustible solution so the wood will be unable to support combustion.

3A1. *Doors*

Because wood doors are such an important part of the cabinet and millwork requirements in an office building and because they receive more concentrated use than almost any other item in this category it is important to describe how they are made so that the architect engineer may be able to make an intelligent choice to suit his requirements. The great majority of wood doors used in an office building are flush panel. They may either be solid core or hollow core. Solid core doors are made in standard thicknesses of 1⅜ inch and 1¾ inch but can be obtained in thicknesses up to 2¼ inches. Standard sizes go up to 4′ × 8′ but larger sizes can be specially ordered. The solid core consists of small pieces (about 2″ × 14″) of kiln dried softwood glued together under heat and pressure. Over this core is glued a cross band of $\frac{1}{16}$-inch-thick hardwood. Hardwood edge bands ¾ inch thick are glued around the four sides and then a $\frac{1}{24}$-inch-thick piece of rotary sliced or a $\frac{1}{28}$-inch-thick piece of sliced veneer is glued to both surfaces. Such solid core doors can be manufactured to obtain a fire rating by using a solid mineral core. The core can be wood particle or flake board[1] for acoustical qualities. There are doors available with lead lining for x-ray rooms. Hollow core doors usually consist of 1¾-inch-wide stiles and 2½-inch-wide rails that are glued and doweled to form a frame. Within this frame is glued a series of slats that are eggcrate shaped, spiral-shaped, or laddered as shown in the diagram. These slats form the core and to them is glued the exterior surface which usually consists of two plies of thin plywood with the third ply on both sides being the finished veneer. A lock block of solid lumber is added to the

---
[1]Flake board is wood flakes or particles bonded with resins under heat and pressure. It is fire-resistant, sound-resistant, and sturdy.

CABINET WORK AND MILLWORK

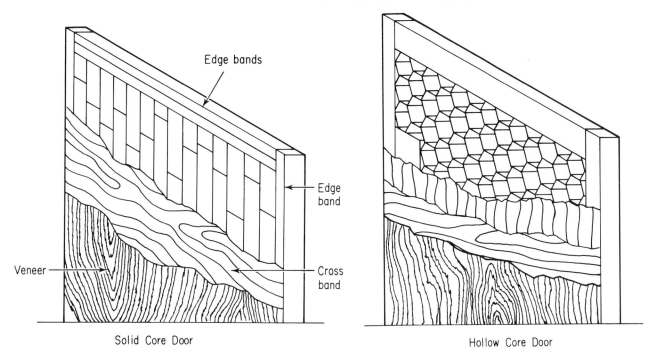

Solid Core Door

Hollow Core Door

core to accommodate the lock and door handle. The well made hollow core door is a good product but the solid core is better, and sturdier, and more soundproof. The solid core is recommended if the budget allows.

3B. Required Workmanship

The quality of the workmanship, especially in such a highly visible material as finished cabinet work is very important and the architect engineer carefully specifies this. The specification calls for the greater part of the work to be assembled in the shop rather than in the field. If work cannot be assembled in the shop, it must be trial fitted before being delivered. All joints are specified to be mortise and tenon, further reinforced by glued blocks and dowels. No work may be installed until the interior plaster or other wet work is thoroughly dry. Wood panels are not allowed to deviate more than $1/32$ inch from the flat as measured from any two points on the panel. All fastenings must be concealed. All joints must be hairline. The finishing is carefully specified. The means by which grounds, nailers, and furring strips are secured to the walls (as by toggle bolts or other fasteners) must have prior approval. The means by which any metal enframement is fastened to the paneling or cabinet work, and its surface treatment, such as the anodized or duranodic finish is mentioned. The architect engineer takes care to see that skilled workmanship is used to enhance the appearance. Such work is usually inspected very carefully after it is delivered to the site, and during, and after its erection.

CONTENTS

1. THE IMPORTANCE OF SCHEDULING
   A. The Bar Graph or Bar Chart
      1. The Private Residence
      2. The Small Business or Residence Building
      3. The Major Office Building

2. THE CRITICAL PATH METHOD

23

SCHEDULING A CONSTRUCTION PROJECT

1. THE IMPORTANCE OF SCHEDULING

One of the most important factors in the construction of a successful project is the proper scheduling of men and materials. When there is an even flow of material as required and the available craftsmen to install it, the entire project is profitable for the owner and the contractors, and enhances the architect engineer's reputation.

1A. The Bar Graph or Bar Chart

In order to plan the progress of a job and to keep track of it during construction, most construction jobs are programmed on a bar graph or chart. This graph, as the diagram shows, is a day by day calendar on which the contractor draws horizontal lines to show when each trade starts and completes its work. The chart shown here is a part of a contract document.

It is a guide so that the owner will know when his building will be completed. Behind the making of this chart there must be vast experience, and a minute study of the plans, specifications, and the site. Although this graph is for a large office building the principles are the same for any structure, with only the materials and their sequence differing.

SCHEDULING A CONSTRUCTION PROJECT

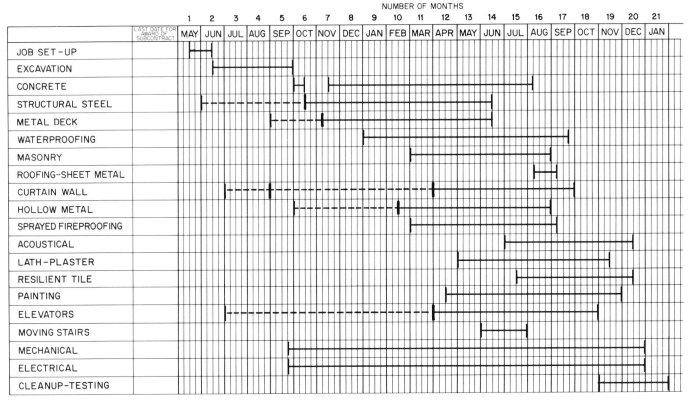

Typical Bar Graph Forecast of Construction Progress.

Any general contractor when bidding or negotiating a job, is in touch with his subcontractors who are estimating the different trades, and in most instances the subcontractors have all been asked for their suggestions for savings in the use of labor or materials. By the time the job is contracted, the builder and the architect engineer should have a very good idea of the availability of labor, material, and the lead time required not only for special equipment or material, but for all heavy equipment. It is the duty of the builder and the architect engineer to call the owner's attention to any possible undue delay so that substitutions may be made or measures taken to minimize financial hardship.

1A1. *The Private Residence*

The successful builder, in cooperation with the architect engineer, would do well to schedule his operation, even for a private residence. It can make a big difference in his own profit margin as well as satisfying the owner.

SCHEDULING A CONSTRUCTION PROJECT

After studying the plans, specifications, and the site, the contractor will know the nature of the soil, any drainage problems, and the best way to get equipment and material to the building site. He should draw up a time schedule and a site plan to show where material is to be unloaded and stored, where the power, water, and sewage lines are to be located, and so on. A typical time schedule would be as follows (a private residence):

January 1. Bulldozer on job to level and pack earth where driveway is to go. Is gravel or stone needed for driveway base? This will prevent the miring of material trucks in bad weather. Bulldozer to scrape top soil from building site and pile it. Have surveyors laid out corners? Erect batter boards at corners.

January 3. Shovel dozer on job to dig cellar hole. Have two dump trucks always available to haul excavated material. Order concrete blocks for foundation wall and form lumber; concrete and reinforcing rods for footings. Are form carpenters available? Are masons available? Prepare framing lumber list and check for lead time. Is all required lumber on hand? Are windows and doors on hand? If all dimensional lumber that is required is not readily available, check with architect for possible available substitutions.

January 10. Have laborers on job to dig footings. Have carpenters on job to build footing forms. Install reinforcing rods and pour concrete for footings.

January 14. (Note 1: This can start sooner if possible) Start digging trenches for sewer and water lines. Does plumbing subcontractor have cast iron soil lines and copper water lines available? Have arrangements been made and permits obtained to connect to water and sewer lines? Check with plumber for lead time if any, on piping, valves, furnace, water tank, toilet fixtures. Check with electrician. Has he made arrangements for power?

January 15. Start masons on laying cellar walls. Does the specification call for waterproofing or drain tile at the footings? Are carpenters available to start framing?

January 25. Start framing. Check for lead time on roofing material. Have plumbing and heating contractor available to check framing details for such work. (This will avoid a great deal of later chopping.) Check availability of mason and material for fireplaces and chimney flue.

January 30. Start mason on fireplaces and flues.

February 7. Start on roofing and temporary waterproofing. Start installation of windows and install temporary doors. Check with electrician to have temporary power brought into house and to start on permanent installation as soon as house is weathertight. Bring in temporary heat. Start plumber on piping.

This schedule can continue through the lathing, plastering, wood trim, tiling, finished floors, etc., but the portion given here illustrates what should be done. It shows that with careful planning, a reasonably sized private residence can be made weathertight and warm in about 6 weeks. If the architect engineer has made all his decisions as to kinds and types of materials to be used and has worked out a careful schedule with the contractor, there is no reason why a house cannot be successfully completed in 4 months or less and make everybody happy, including the owner.

The architect engineer and builder should be particularly wary of certain materials that require special colors or that are in other particulars not "off the shelf." Special runs of ceramic tiles, or kitchen cabinets, or bath fixtures may take months to get. There may be a short supply of tile setters or floor layers. These men must be notified ahead of time and given accurate dates. The writer knows of almost completed houses that had to stand idle for many weeks waiting for certain craftsmen or special materials. Meanwhile mortgage interest, taxes, and overhead go on and the unfortunate owner may have to find an interim place to live. Successful development builders have worked out scheduling to a fine point. They have crews of specialists who follow one another very closely. One crew builds foundation forms; another crew sets reinforcing; a third crew pours the concrete, and so on. There are flooring crews, and framing crews, and roofing crews. This close scheduling is one reason that development houses can be built more cheaply than "tailor made" houses. There is no reason, however, why the private builder cannot take advantage of some of their techniques.

1A2. *The Small Business or Residence Building*

In the case of the small business or residential building the technique of scheduling is the same, although, of course, the problems are different. As in the case of the private residence there must be a team effort between the architect engineer and the builder. They must be able to identify the hard-to-get materials and with the approval of the owner be prepared to make substitutions. For instance, if bar joists are hard to get at the scheduled time or are not available in the sizes required, there may have to be a recalculation for another size, or a substitution made to a precast concrete floor system or a cast-in-place floor system. A typical detail of a schedule for such a building would be as follows:

> *January 1.* Start site clearing and preparation for material storage, construction office, and other site facilities. Start survey crew to lay out exact outline of structure. Order reinforcing and form lumber for basement walls and arrange for either batch mixing on site or transit mix concrete. Are

form carpenters and lathers for reinforcing available? Is there a water situation below grade? If so, are there pumps available? Is there "ledge rock"—or large boulders? Is a skilled blaster available? Before arranging for removal of excavated materials, is there a low spot on the site where it can be used for fill?

January 6. Start bulldozer or power shovel to excavate for cellar and footings for foundation walls. Determine that there are sufficient dump trucks available and check the distance from the site to the dumping facility. There should be enough trucks to always have one on the site available for loading (power equipment and truck rental is very expensive and the equipment should be used efficiently). Check with plumber regarding sewer and water permits and material, and for temporary water. Check with the electrical contractor for temporary power supply. Are the necessary transformers available? Are extra power poles necessary? At this time every kind of material and equipment to be used in the entire operation should be carefully scheduled and a job meeting should be held with representatives of every trade except possibly the finish trades (plastering, painting, finish floors, etc.) At this meeting every trade representative should be questioned as to the availability of men and material when they will be needed.

January 20. Start hand labor on footings and excavation, and carpenters and lathers on forms and reinforcing. Arrange for enough transit mix trucks to maintain a continuous pour. Double check that all chases and holes through walls are in correct places before pouring. If a local building department inspection of concrete or footings is required, notify the department and arrange for such inspection so as not to delay the pour. Check site to see that trucks can reach as much of the walls as possible with a chute instead of using hand labor. Are temporary power transformers on job? Is temporary water now available?

February 1. If the structure is to be of reinforced concrete—start form carpenters on interior column work as exterior wall is being completed. If structural steel is to be used, the first tier of columns and beams should be on the job. (It is assumed that such small buildings are not in the centers of cities where storage on the site is a problem.)

1A3. *The Major Office Building*

The intricate scheduling of the construction of any major structure requires expert knowledge, a thorough study of the plans and specifications, and a detailed knowledge of the below-grade conditions as well as constant meetings with subcontractors and material and equipment manufacturers. The completion within the set time schedule of such a structure involves large amounts of money. There may be lease dates or a manufacturing schedule to be met, and the carrying charges for interest on the financing during construction as well as taxes, and overhead can be very heavy indeed.

During the course of the negotiations, the general contractor, the architect engineer, and the owner have been meeting frequently. The general contractor has been talking to his subcontractors and has been questioning the architect engineer on possible changes in plans or specifications. During the bidding the general contractor and his subcontractors have been in possession of very precise lists of acceptable material and equipment, and have therefore been in touch with the manufacturer or suppliers of such material to check availability and time schedules.

Sometimes it is possible for the owner and the architect engineer to preorder certain equipment such as structural steel or other material that requires long lead time, or to contract for the excavation and foundation before the general contract is awarded. In such cases the general contractor takes over these contracts as part of his responsibility, and valuable time has been saved. It is possible in many communities to obtain permits to excavate without filing complete plans. Structural steel requires about 6 months between contract award and first delivery. Certain heavy equipment can require more than 6 months. When the general contract has been awarded, the first step for the architect engineer, contractor, and owner, is to meet and come to a mutual understanding of the owner's requirements for a certain completion date, the planner's requirements for quality and performance, and the builder's ability to comply with these requirements. From these and other meetings and communications the construction schedule is born.

The bar chart on page 312 was drawn for a building that has been completed. It was completed ahead of schedule, which is a rarity in the construction business. The building was able to accommodate 1 tenant 2 months before it was fully completed and thereby saved him almost $100,000 in the premium rentals he would have to pay had he remained at his former location. The good will in such a case is beyond price. It was a financial success for the general contractor, his subcontractors, and the owner. It has enhanced the architect engineer's reputation for careful and thorough planning and supervision.

It will be noted that the construction started on May 15 and the preliminary site planning and layout were scheduled for a month. During this time a site layout was made for trucking facilities, material storage, and offices for the general contractors, subcontractors, and the architect engineer's and owner's representatives. Temporary water and electricity were brought into the site. Arrangements were made for sanitary facilities.

Survey crews started layout of the site and established bench marks (points of reference). The excavation, shoring, and dewatering started on June 15, before the site planning was completed. In this case, because of anticipated trouble with the ground water and the poor bearing capacity of the soil, a geologist and soil expert was retained by the architect engineer to lay out the locations and the number of the test borings required to give a thorough knowledge of the below-grade condition. Some borings were arranged with a hollow perforated casing so that the flow of ground water could be measured. From the knowledge gained from the borings the type of shoring for the perimeter earth banks was determined and the excavation subcontractor was alerted. The excavation was 200′ × 400′ and was 50′ deep.

SCHEDULING A CONSTRUCTION PROJECT

Approximately 150,000 cubic yards of earth was excavated and removed by a continuous line of dump trucks in just over 12 weeks. The perimeter bank shoring followed the excavation as it went down. Within 2 months—Aug. 15—of the start, the general excavation had gone down to bottom grade at one end of the site and excavation could start for pier holes and footings for foundation walls. Because there had been a fairly dry spell for several months before, the expected sheet of underlying water was found to be almost nonexistent. The point is, though, that pumps and well points were immediately available if they became necessary.

As the pier holes and footings were being excavated, lumber and reinforcing rods were delivered. Arrangements were made for city inspection of each hole as it was completed. A careful schedule had been worked out with the supplier of the transit mix concrete which was to be used for the piers, footings, and walls. The running time of his trucks was agreed upon and he set aside enough trucks to always have one in the process of mixing on the site as the previous one started to unload. Because of thorough site preparation all the concrete was poured into place directly and none had to be hand-hauled at prohibitive labor rates. Arrangements had also been made to have an inspector at the mixing or batch plant when necessary.

On October 15, 5 months after the start of the project, the first structural steel was delivered. It had been ordered by the general contractor immediately after he had been awarded the job. He had been checking constantly on the progress of fabrication. The architect engineer had done his part by the expeditious checking of the steel contractor's shop drawings and the following through of changes. Erection of the steel was started immediately upon its delivery.

The structural floor had been scheduled and started to arrive when steel was up two floors. It could therefore be used to cover the open floors instead of using temporary planking. Each material arrived when it was due and there were workers on hand to install it. Each job is different but the thorough knowledge and careful planning for each is the same. There were, however, some circumstances in this job which deserve special mention. The first is that the air conditioning plant and the fans for the upper floors were to be placed on the roof. While this is becoming fairly common practice it meant that the complicated and exacting work of setting and connecting the compressors, chillers, and condensers could not be started until the steel (the same goes for reinforced concrete) was topped out. This meant the middle of June before this work could be started, or more than a year after the project was started. It was also required that all the equipment be assembled and ready to be hoisted to its location while the structural steel cranes were still hoisting the last of the structural steel, and that the heavy concrete pads on which such equipment was to rest had to be poured and ready. It also required the air conditioning contractor to have his vertical main ducts in place almost as the structure went up. Another special consideration was the precast concrete mullions mentioned in Chapter 12. The scheduled start for their erection was December 1 or approximately 6 weeks after structural steel was started. In order to maintain a constant supply of these mullions to follow the steel progress many things had to be determined. The fabricating capacity of the manufacturer was studied and agreed upon; the curing time after manufacture was determined; samples of varying mixes were inspected and the final mix was chosen. Although a single mullion could be cast and ready for erection in less than 5 weeks the manufacturer was ready to start fabrication 5 months before any mullions were required. By the time they were required he had several hundred on hand. His trucking time was determined and sufficient trucks were made available to always have one waiting. The result was a smooth running operation exactly on schedule. This illustration indicates only a very small part of the skill and knowledge that is required to successfully complete a construction job on schedule. The morale and spirit on a job where everything is on time and ready to go is beyond price.

It may be of interest here to mention the scheduling of structural steel on another construction job—a 42-story building requiring 25,000 tons of steel. The building was located at a very busy avenue and two heavily traveled cross streets. The steel trucks had to come one at a time, unload and leave, with one coming just as the other was leaving. The steel was unloaded from freight cars at a railroad yard about 2 miles from the site. It was placed on flat bed trailer trucks in the order of its erection and held in the yard. As one truckload of steel arrived at the site, the next one was started from the yard by telephone orders. With normal traffic and unloading time it arrived on the job just as the preceding truck finished its unloading. This constant, almost assembly-line process was possible because the architect engineer, owner, and contractor had prepared for it months before. The shop drawings had been approved on time and the steel mill and fabricators had been constantly checked. The scheduling and construction of a high-rise office building is unlike any other type of construction. In a tall structure the bottom floors may be enclosed, being plastered and finished while steel is still being erected on top, and concrete floors are being poured just below. The separation of the delivery entrances for the various materials and the vertical transportation of the different crafts to their separate destinations require special experience.

One other very important consideration in progress scheduling is the available supply of labor. The architect engineer, contractor, and owner must be fully familiar with the labor situation in the locality. If certain labor contracts such as steel or aluminum are due to expire during the course of the project, preparation should be made to obtain and store the affected material before the expiration of such contract. If field labor is involved a substitute schedule should be designed to work around this trade if possible. If the supply of labor in certain trades is uncertain it may be possible to substitute other material or methods such as the spray fireproofing of steel beams in-

SCHEDULING A CONSTRUCTION PROJECT

stead of poured concrete if skilled form carpenters are not available. One serious cause of delay could be jurisdictional disputes between crafts as to which installs certain material. Some cities are notorious for this kind of labor trouble. It may be possible to have representatives of all the crafts meet and decide which trade installs what from the plans and specifications before the job starts. In this field there is no substitute for careful, advance planning.

2. THE CRITICAL PATH METHOD

The Critical Path Method (CPM) of building construction scheduling is a graphic representation of the entire construction process. It shows when every kind of material is to be on the job, and goes back beyond this to show when such material must be ordered, and goes back beyond this to show when shop drawings must be approved. It shows how one trade interlocks with another. As an example, if a sewer line has to be laid at a certain date the CPM goes back to the surveyor staking out the line and grades; the plumber ordering the necessary sewer pipe; and the excavator digging the trench. It also mentions that a permit to connect to a city sewer must be obtained. The bar chart on the other hand is a line presentation that very simply states

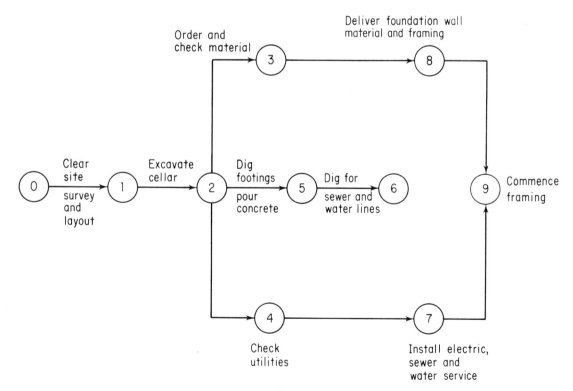

A Small Portion of a Construction Job.

319

when every operation is to be started and completed. The intimate knowledge and planning of the job—the work that goes to produce a workable bar chart is not shown. The CPM does show this. It forces the architect engineer and contractor to meet and plan the entire operation step by step. Every step of construction and when to order each piece of equipment in order to get it to the job on time is shown on the chart. It forces decisions, or at least forces the entering on the diagram of the best available information in order to complete all the paths. The CPM diagram shown here is an example of a very small portion of a construction job. It shows how the order of construction can be diagrammed and the paths from one process to another are numbered. If a date is missing or not immediately available the planner must enter a date that will complete a process at the required time. He is then forced or urged by the diagram itself to obtain the material or information on the date he has entered.

For instance, if the diagram shows that the structure will be ready for elevator motors at a certain date 6 months in the future and the lead time for such motors is 5½ months—all concerned are made aware of the fact that they had better make up their minds and award an elevator contract. The critical path method can be programmed and computerized. When it was originally developed around 1957, it was thought that its insistence on careful planning would appeal to many contractors, architect engineers, and owners. So far this has not been the case. Not many owners or contractors are willing to pay the cost for the careful planning that is necessary. The capable contractors feel that they can do this work and do it without the CPM pushing them into it. The lesser contractors use the "seat of the pants" method which often works. When it doesn't, the result justifies the common complaint about the high cost of construction.

24

DESIGNING FOR THE HANDICAPPED

All buildings that are likely to be used by the general public should have provisions for the physically handicapped. Many of these provisions are quite simple and inexpensive, and only require a little forethought on the part of the architect engineer as he is designing the building. The resultant good will and a sense of having discharged a responsibility toward our less fortunate neighbors is worth the effort it takes.

There are many barriers to the use of a normally designed building by wheel chair patients or the blind. The first obstacle is access to the building. Flights of stairs leading to a front entrance or doors that are not wide enough to permit the passage of a wheel chair should be supplemented by a secondary means of access from the street level. A ramp with a grade of not more than 8.3% leading to an elevator or the front door should be provided. The ramp should have a handrail on one side if possible and be nonslip. Ramps and other means of access without the use of stairs can be used by wheel chair users as well as people with heart conditions, who now number over 5 million in this country.

Ramps as a means of ingress and egress are recognized by city codes and following are the requirements of such a code:

1. No ramp can be over 30 feet long.
2. If a ramp is sloped more than 1 in 10 it must have a nonslip surface. All exterior ramps must have nonslip surfaces.
3. A ramp may change direction only at a level platform.

DESIGNING FOR THE HANDICAPPED

    4. Ramps must end at a level platform at top and bottom. If a door swings outward to the platform such platform must be at least 5 feet wide.

    5. No door shall open onto a sloping portion of a ramp.

Easy access is of particular importance in medical buildings. In a previous chapter on site planning the author mentioned a medical building. This building is located on a knoll and the parking lot is very poorly arranged and so narrow that cars have to park between rows of columns which support the upper portion of the building. It is very apparent that the knoll is there because the owner of the building did not wish to pay for the small amount of rock blasting necessary to remove it. There are flights of steps leading to the front or rear entrances. The building is occupied by several orthopedic physicians. It is inaccessible to the handicapped and in time when other space becomes available it will be vacated.

Entrance doors should have a clear width of at least 32 inches and be operable by a single effort. By law, entrance doors open outward, so a platform wide enough for a wheel chair to clear the door as it opens should be provided. The planner must provide ramps, street level entrances, and wide doors for goods—why not also for people?

In buildings where the handicapped are likely to use stairways, overhanging nosings should be avoided and the risers limited to 7 inches. They should be provided with handrails at a height of 32 inches.

Wheel chair users require special provisions in toilet rooms. One booth should be at least 3' wide by 4'8" deep and with an outward opening door. It should have handrails on either side of the bowl. If possible provision should be made for an accessible drinking fountain or a cup dispenser within reach near a cold water source. Fountains 36 inches high are now made which can be reached by everyone. The Telephone Company can supply public telephones for both wheel chair users and the hard of hearing. These should be suitably marked and can also be used by the general public.

To prevent the blind from walking into stairwells or other dangerous places many buildings have the knobs or handles on doors leading to such hazards roughened or knurled.

The signals on elevators at landings should be, and as a matter of fact almost always are, audible as well as visible. This is for the benefit of the general public but it helps the deaf and the blind immeasurably. In buildings that are principally used by blind people the elevator emits an audible signal that differs for up and down.

This brief chapter can be aptly completed by a quotation from a handicapped citizen, Charles E. Caniff:

> "For those of us with limited mobility, architectural barriers prevent free access to those buildings which we must enter to work, to vote, to worship, to learn, to play, even to buy a stamp. To fulfill our responsibilities as

citizens, we often must circumvent these barriers by entering through the rear door, where freight is hauled in and garbage hauled out, and make our way through coal bins, storerooms, and boiler rooms to reach a freight elevator which can accommodate our wheelchairs. Architectural barriers have made us 'back door' citizens."

CONTENTS

1. DISCUSSION OF ITS IMPORTANCE
   A. Safety Meetings

2. PROTECTION OF THE PUBLIC
   A. Sidewalk Sheds and Fences
   B. General Protective Measures

3. PROTECTION OF WORKERS
   A. Hats and Safety Belts
   B. Protection of Openings
   C. Rubbish Removal
   D. Vertical Transportation
   E. Scaffolds
   F. Projectile Tools

4. PROTECTION OF EXCAVATION, CONCRETE, AND STEEL CONSTRUCTION

5. FIRE PROTECTION
   A. Standpipes and Fire Extinguishers
   B. Housekeeping
   C. Fire Watch
   D. Vertical Transportation for Firemen

25

SAFETY OF THE PUBLIC, WORKERS, AND PROPERTY DURING CONSTRUCTION

1. DISCUSSION OF ITS IMPORTANCE

This chapter has a long title and concerns a subject which should be looked into very carefully by the architect engineer, the owner, and the contractor. Safety should be a matter of contractual obligation, and provision must be made for constant supervision to check that such obligations are met. This is a matter which affects the morale of a job (many accidents make workmen jumpy and unproductive); money (rates for fire and casualty insurance are high) and losses are not always fully compensated and caution helps prevent the waste and sorrow of human injury. Public authority has made the safety of people and property a subject of code regulations that are strictly enforced. It is imperative that the owner and all the contractors be concerned with safety, if only to control insurance rates (a general contractor or subcontractor with a poor accident record may find his insurance rates so high that he can not afford to bid as low as his competition). Safety must be the concern of the architect engineer who can exert considerable influence in creating a safe job through precise specifications, discussion with the contractors at job meetings, and regular inspection of the job.

1A. Safety Meetings

During the course of any important construction job there should be several meetings devoted to safety. These should be attended by representatives of all subcontractors on the job as well as the insurance carrier and architect engineer representative. The meeting should be conducted by the general contractor and an agenda and minutes should be prepared. It is surprising how a well run meeting can establish the duties and obligations of all concerned. It also serves as a constant safety reminder. Insurance companies and public authorities issue posters on safety. These should be displayed at water fountains and other places where workers are apt to congregate.

2. PROTECTION OF THE PUBLIC

The first concern of public authority as well as of the architect engineer, contractor or owner, is the safety of the public. The innocent bystander or passerby in most cases knows nothing about construction job safety. He must be protected against falling objects, tripping hazards, projections that may tear his clothes (or him), and the danger of being run over by vehicles going in and out of the project.

2A. SIDEWALK SHEDS AND FENCES

In any location where a construction project abuts on a publicly traveled street all specifications and codes require a substantially constructed fence to keep the public out of the site. Generally in cases where a building is closer to a public street than one-half its height, a sidewalk shed must be built to protect the public against falling objects. Some codes and specifications, especially in large cities, are explicit about its construction. They call for watertight heavy plank shed roofs capable of bearing at least 150 pounds per square foot. The underside of the shed above the sidewalk must be well lighted. The building side of the shed must be enclosed with a solid fence to keep debris from falling or bouncing onto the walkway. The sidewalk itself must be smooth and free of tripping hazards. Some owners look on the sidewalk shed as a means of advertising and gaining the good will of the public. Many of us have seen such sheds lined with posters, with recorded music, and with pick-up telephones that tell about the company or the progress of the job. Some companies use the outer side facing the street to show a picture of the building and advertise the owner's business. Many are used for holiday season displays. If the shed has to be there, why not use it and make it attractive?

2B. GENERAL PROTECTIVE MEASURES

In addition to the sidewalk shed (it is also called a sidewalk bridge), codes and contracts call for the protection of temporary electric boxes, temporary water lines, or moving machinery from contact with the public. In addition it is the obligation of the contractor to maintain fire preventive, sanitary, and other public facilities in full working order during temporary connections and, of course, after the permanent connection.

At any construction job there is always constant crossing and re-crossing of sidewalks and public roads by vehicles carrying construction materials. Trucks carrying excavated material, steel, or concrete, or other construction material must be directed in and out with minimum delay and with complete protection to the public. Contracts call for flagmen to warn the walking public or oncoming traffic of such truck crossings. One can sometimes see a pedestrian walking right into the flagman who is trying to keep him from walking into a truck crossing the sidewalk.

3. PROTECTION OF WORKERS

3A. Hats and Safety Belts

The construction worker is always furnished with a hard hat, and often with safety goggles, and a safety belt, and it takes constant vigilance to get him to use them. A hard hat should be furnished to every one, worker or supervisor on the job, and there should be an absolute prohibition against any one entering a construction site without such a hat. Safety belts and goggles are another matter. Only constant reminding will ensure their use. There was the case of a workman who was standing on a rolling scaffold inside a building and leaning through a window opening to adjust a gasket prior to glazing. His safety belt, which was supposed to be attached to the scaffold, was not. Any distraction or the slightest slip would have resulted in a 30-story fall. When he came into the building he was told to put his belt on. He did it grudgingly. Several hours later he was observed doing the same thing. He was paid for the day and discharged, and his union business agent concurred.

3B. Protection of Openings

A building under construction has many openings in the floors for pipe shafts, stairwells, elevator shafts, and other utilities. The outer edge of the building is open. Workers who are carrying, connecting, and erecting things and are busy on the job must be protected against falls down these openings, so the contractor must place barriers around them strong enough to hold a man. They should have protective boards (toe boards) at the floor as well as at waist height. If possible, such openings should be planked over until they are used. The outer edge of a building can be called to the workers' attention and protected by the use of a steel cable firmly fastened to the per-

SAFETY DURING CONSTRUCTION

imeter column. It can be made part of the contract of the structural steel or concrete work. Protection must also be given against hoisting cables that sometimes run through a building.

3C. Rubbish Removal

Another important precaution against accidents is the removal of rubbish. The floors of a building under construction can present an incredible picture of clutter. Empty packing cases, pieces of pipe, tangled wire, masonry, and plaster, and wood must be removed daily if the job is not to choke on itself. Rubbish also masks other hazards. The general conditions of a contract always call for rubbish removal at short intervals. Many contractors use large hollow metal rubbish chutes with openings at each floor and a hopper at the bottom for direct loading into a truck. In low buildings such chutes may be built of heavy planking.

3D. Vertical Transportation

As a construction job progresses upward, means must be provided to get workers and material to the scene. Material is lifted on bucket hoists or platform hoists and some material, notably concrete and steel, is lifted by derricks. All of these devices require expert rigging. Some codes call for riggers and hoisting engineers for cranes and hoists to be licensed. The cables and fastenings must be carefully inspected. The hoist towers must be thoroughly enclosed when they run through a building. If they are installed on the outside of a building, especially a tall one, a competent engineer should examine the strength of the structure, its guying and the manner in which the hoist tower is fastened to the building. It must be completely enclosed. The runways to the tower must be railed and the doors to the openings sturdy and capable of being securely closed. Only experienced signalmen should be allowed to do the signaling. This may sound incredible but it has been done, and no doubt will be done again—a worker sticking his head into a shaft to see where the hoist is! A sturdy door, with a good closing device, may give him pause. Bucket hoists are used for concrete and they run in exterior towers or they are lifted by cranes. They must be carefully rigged, and enclosed, and guided by experienced signalmen. In structural steel or concrete structures, stairways follow very closely behind the superstructure. As the building goes higher, however, the workers must be transported mechanically and safely. Workers are forbidden to use material hoists but there are temporary elevators available that can meet safety standards. They are counterweighted and have brakes and other safety devices. Such elevators usually follow behind the structure by several floors. The upper floors of a building under construction, however, can usually be reached only by ladders. This is the case also for pits, excavations, and low subsidiary structures. Such ladders are the subject of definite published rules by state or local labor safety codes, for safe and sturdy construction.

3E. Scaffolds

Scaffolding is a very necessary aid in all construction. When carpenters nail sheathing on the outside of a private residence they use scaffolding, and when a painter does your ceiling and places a plank between two ladders, he is using scaffolding. Scaffolding of various heights and materials is used on every floor of a building during construction, and pole scaffolding can be built up the side of a building for many stories. Hanging scaffolds are used by glaziers, masons, and curtain wall workmen. All of these scaffolds are subject to the requirements of building codes, insurance carriers, and labor laws. It is suggested that the architect engineer and owner be aware of these requirements.

Requirements specify scaffolding according to the duty it is to perform. A painter's scaffold can be much lighter than one for a stone mason. The allowable width of platforms, the clear span of planking, and the size of the uprights or poles can all be found in tables. There are many scaffolding devices now made that may be wheeled about and raised and lowered as necessary. These are good for specific jobs but when an entire lobby ceiling has to be plastered there is no substitute for a planked-over scaffolding on which men can walk.

Suspended scaffolds, which are potentially the most dangerous, must be tested carefully before use. A usual test is to load the scaffold to four times the load of men and material to be placed on it. The scaffolds must be provided with guard rails at a height of 3'0" to 3'6" and must also have a toe board so that a man cannot slip nor can material fall under the guard rail.

3F. Projectile Tools

The projectile tool is really a hand gun that shoots a bolt or other fastening into a concrete slab or even into a steel member. The force of the projectile is considerable and it can do at least as much damage as a bullet. The use of such tools is therefore strictly limited by code and labor law. The tool must have a safety feature that prevents its discharge if it is dropped. It must only discharge when the mechanism is deliberately activated. It must never be left unattended. It must be used only by the authorized agent of the

SAFETY DURING CONSTRUCTION

owner of the tool. There is at least one case where the careless use of such a tool shot a bolt across a street and through a glass window. The projectile tool is a dangerous weapon and should be treated accordingly.

4. PROTECTION OF EXCAVATION, CONCRETE, AND STEEL CONSTRUCTION

This section will concern itself with the protection of property. The resultant safe practice also protects the public and the construction worker. Specifications and codes call for the protection of adjoining property. If the new construction goes down below the foundations of an adjoining property this property must be protected by a retaining wall and underpinning. A party wall[1] must be carefully braced and repaired when necessary. If a tall building is built alongside a low one, the roof of the low building must be protected.

Excavations must be protected against the collapse of earth. Generally, the shoring and bracing of the hole is required when it goes down more than 5 feet. Large excavations must be sheet piled. In most cases the architect engineer has either designed or approved the design for such shoring, bracing, and sheet piling. But where he has not, there are codes and tables specifying the size of the sheet piling, stringers, and bracing for various types of soils. The possibility of law suits or construction delays should alert the owner to the importance of getting this work done.

The protection and shoring of concrete during the time it is *being poured and until it attains its full strength* should be carefully supervised by the architect engineer and always is carefully supervised by the public authority. Many lives have been lost through the collapse of concrete structures during construction. The forms in which concrete is poured must be very well built and well braced laterally as well as from below. It should be remembered that wet concrete weighs about 150 pounds per cubic foot and can exert tremendous lateral pressure. This pressure, plus its dead weight, must be held up until it attains its initial strength (after 7 days), and then it must be held up by reshoring until it gains its full strength (after 28 days). This shoring, and bracing, and reshoring should be done under the direction of a competent engineer.

The hoisting and erection of structural steel is a dangerous operation, but in the course of many years' experience it has been found that the majority of steel workers are thoroughly grounded in safety rules and practices. The great steel companies have played a major part in this matter. This, plus constant safety meetings, and alert engineers and foremen, account for an

---

[1] A party wall is a common wall between two buildings. If one building is torn down the party wall must, of course, remain.

excellent safety record. In the erection of steel, as the members are fitted together, they are held by a single temporary bolt hand-tightened by a wrench until the final bolting, riveting, or welding is done. Care must be exercised that this temporary bolting does not get too far ahead of the permanent fastening. Structures have been known to collapse. In smaller structures or in any case where open web joists are used, the steel members should be permanently fastened almost as soon as they are placed. Until they are, no load or planking should be placed on them. During steel erection the upper deck on which the men are working can be planked to some extent but the deck two floors below the top should be completely planked or covered with steel decking as soon as possible.

5. FIRE PROTECTION

A building under construction is probably more vulnerable to fire than it ever again will be, and a fire in a building under construction can be disastrous in time, money, and morale. Materials come in boxes and packing cases that are discarded; carpenters cut pieces from the ends of planks and other lumber; there is oxyacetylene, or electric welding, or cutting being done; plumbers are heating lead; workmen are smoking; there are live electric cables snaked across wet floors. If it is a concrete building or a steel building with concrete floors there is a forest of form lumber, braces, shores, and reshores. It can readily be seen that extraordinary precautions must be taken to prevent fire. The architect engineer and the public authority have very strict specifications and rules regarding fire-preventive measures.

5A. STANDPIPES AND FIRE EXTINGUISHERS

Most codes and specifications call for temporary standpipes or the permanent standpipes to be installed if the building is to be over 75 feet high. The standpipes, if permanent, should be of the size and in the locations called for by the fire protection code. They must have a hose connection at each floor and be connected to a siamese connection at the street so that fire department pumpers can pump water up to the floors. The standpipe must be located so that a 20-foot stream from the end of a 125-foot hose can reach every part of the building. Some architect engineers specify a temporary standby fire pump connected to the building's water supply to pump water up into the building in case of fire until the fire department arrives. They also specify fire extinguishers at readily accessible places on each floor. These should be checked daily.

SAFETY DURING CONSTRUCTION

5B. Housekeeping

Good housekeeping on a job is the contractor's best ally against fire. It involves the daily removal of rubbish and the piling of materials to leave uncluttered aisles. All contractors and subcontractors have job offices where they keep their plans and perform their paper work. These construction shanties are usually built on the lower floors as soon as the concrete is poured. The architect-engineer should insist in his general conditions, under fire protection, that these job offices be built of nonflammable material like transite, sheet rock, or metal. Many fires have started in such wood shanties.

5C. Fire Watch

The wise owner or architect engineer will insist on a fire watch during the period that crated material is arriving and concrete is being poured. At this time the entire operation is very vulnerable to fire. Watchmen should make their rounds constantly during the entire time that regular workmen are not on the job. The watchman should be notified if welding, cutting, or other similar operations have been going on during the work day so he can be especially vigilant at such locations. In several jobs recently supervised by the author, the watchman either carried a walkie-talkie connected with a man who was patrolling the street perimeter of the building; or he could reach a button on each floor which actuated a loud gong at street level to warn the downstairs man to call the fire department. The cost over and above the normal watchman service? $30,000 on a $30,000,000 building.

5D. Vertical Transportation for Firemen

One other essential safeguard against fire in tall buildings is the installation of the standby elevator when the building goes over 100 feet in height. This can either be one of the temporary elevators that has been installed in the building to carry workmen to the upper floors or a construction elevator suitable for carrying people. This elevator must be manned at all nonworking hours so that firemen can be taken to the scene of the fire without delay. In one large city, the local fire department pays frequent visits to all large buildings under construction. Key firemen in each district know where the elevators are, where the standpipes and siamese connections are located, and are familiar with the general building layout. It is hoped that more cities will adopt this excellent cooperation.

The safe building has an atmosphere about it that can't be missed by the experienced eye. Workers move about with confidence through uncluttered floors and around carefully enclosed floor openings. Materials are stored where they can be reached easily. Fire extinguishers are in well marked locations. The general feeling of efficiency and on-time performance reflects itself in many dollars and much satisfaction for everyone concerned with the operation.

CONTENTS

1. THE DUTIES AND RESPONSIBILITIES OF THE ARCHITECT ENGINEER
    A. Proper Inspection During Construction
    B. Forward Planning. Allowance to be Made for Future Use
    C. The Setting up of a Maintenance Program

26

AT THE COMPLETION OF CONSTRUCTION HOW SOME FUTURE PROBLEMS MAY BE AVOIDED

1. THE DUTIES AND RESPONSIBILITIES OF THE ARCHITECT ENGINEER

The architect engineer's duties and responsibilities do not terminate when construction is completed. There is very rarely a time when he is not called back by the owner to advise on unforeseen problems of maintenance that arise in even the best designed and constructed building. Some future problems can be avoided by careful inspection during construction to make sure that the plans and specifications are followed. Many problems can be avoided by a degree of foresight on the part of the owner and architect engineer and by the setting up of a program of preventive maintenance by the owner with the advice of the architect engineer.

1A. Proper Inspection During Construction

We will discuss these in order, the first cause of future trouble being the lack of adequate inspection or just a slip-up in inspection during construction. Of course, no inspector can be everywhere at once; the architect engineer must therefore set up a list of inspection priorities. As buildings are built for protection from the weather, one of the first priorities is to keep weather out of the building. This means careful inspection of roofing, flashing, exterior wall structure, window openings, caulking, and any exterior struc-

AVOIDANCE OF FUTURE PROBLEMS

tures subject to weather such as entrance porticos, penthouses, skylights, elevator bulkheads, and so on. All of these pieces of work cannot be looked at constantly as they are being built but there are certain times when they should be inspected. For instance, once the inspector is satisfied that the roofer is using the proper amount of pitch, the proper weight of roofing felt and is overlapping properly, he can turn his attention to other matters, but he must watch carefully to see that the flashing is properly soldered and is overlapped in the specified amount by the roofing. He must see to it that expansion joints are placed where they are shown on the plan. The caulking of all joints in the exterior skin must be constantly spot checked. If the exterior wall is of masonry the inspector must see to it that all work is fully bedded in mortar and that the back of the exterior masonry is fully slushed with mortar. The glazing is very important. A vulnerable spot for water leakage is between glass and frame, especially in a tall structure which can sway in high winds. The neoprene gasketing and the neoprene blocks which keep the glass from contact with the metal frame must be checked. This can prevent water leakage as well as the possible cracking of the glass.

Another cause of trouble is noise. The specifications are explicit about all the precautions the contractor must take to avoid sound transmission by air leakage or by vibration. In spite of this, a careless workman can undo the most carefully laid plans. There are instances where transmitted vibrations have made an entire office unhappy and where it took dozens of hours to find the cause. In one case it was a water line lying directly on a steel beam and transmitting the pump vibration to the steel structure. In another case someone had forgotten a sound attenuater lining in a duct. In a case of air-transmitted sound the sound barrier between two offices in a perimeter air conditioning enclosure had been left out and one salesman was listening to another salesman's conversations. Not all of these troubles could have been eliminated but inspection of sensitive areas could have stopped some of them.

Cracks in walls and ceilings are a cause of annoyance in private homes as well as in large structures. A careful inspection of the underlying soil condition and the foundation itself can avoid much of this annoyance, especially in the private home. In the large structure the architect engineer must make sure that expansion joints are properly installed and that walls are properly anchored and ceilings securely hung.

The continuous inspection of steel and concrete is usually performed by a subcontracting inspection firm which does nothing else, but the architect engineer must be constantly aware of the inspection reports. An example of what can happen through lack of proper inspection was mentioned in Chapter 5 on Foundations. In this case, while large concrete piers were being poured into forms going down to rock, the forms were pulled up faster than the concrete set and this allowed soil to press into the wet concrete and displace it. Apparently no one wondered why less concrete was

AVOIDANCE OF FUTURE PROBLEMS

being used. Had they been curious when it first happened, a great deal of money could have been saved. This is also true for any cast stone that is used. A careful specification for this will be of no avail if it is not followed. Cracked terrazzo floors can be unsightly and very often are caused by improperly placed expansion joints or improper bedding.

All this does not mean that contractors are not trying to do their best but it does mean that careless or incompetent workers can make mistakes. It is recommended that the architect engineer and the project inspector enlist the help of the various contractors' foremen and other supervisory workers to especially check possible future trouble spots. It will be found that almost without exception they are all interested in doing a good job.

1B. Forward Planning. Allowance to be Made for Future Use

Proper forward planning and design on the part of the architect engineer and owner can avert many serious problems, some of which are very costly to correct after the building is completed. The simplest example is the providing of enough headroom in the attic of a peaked roof house for future expansion or bringing in a large enough electric service for unforeseen future needs. Some lack of forethought is ludicrous as in the case of the boiler that needed retubing. It was so close to a masonry wall that the tubes could not be pulled out or new ones put in. The wall had to be practically torn down. Fortunately it was not a bearing wall.

Forward planning will allow for extra electric service in a large building by specifying an empty conduit to be run up through a pipe shaft, or by using a bus duct with spare capacity, or by allowing plenty of extra room in underfloor electric ducts, or by having the plumbing contractor install empty wastes, vents, and water lines in one or more locations in an office building to allow for future toilet rooms or other use of plumbing facilities. The walls of elevator cabs are a constant source of trouble. They are bumped, scratched, and abraded, innocently or otherwise. The material should depend on the type of occupancy that is expected or if a material that can be scratched is used it should come in easily removable panels—otherwise it may cost many thousands of dollars to entirely replace the elevator cab walls.

This book has discussed elevatoring (Chapter 11) and has stressed the need for sufficient elevators to adequately meet a building's requirements, but sometimes there may be a requirement over and above the normal. Perhaps an occupying tenant, with whom negotiations are going on while the building is under construction, has a special requirement for large catalogue mailings or has very heavy interfloor traffic in people or papers. In such a case an alert owner or architect engineer can probably arrange for a heavy-duty dumbwaiter or even an extra elevator (it there is time), and can probably get the tenant to pay for at least part of it.

Requirements for special sound insulation should be designed for and built into a building if the foreseeable future occupancy is likely to call for this. Many architect engineers design extra heavy steel for certain portions of the floor areas to accommodate future requirements for heavy floor loading by computers, or files, or other special tenant equipment.

There are many other examples of the trouble that can be caused by inadequate forethought. In a large building that was going to have its sealed, fixed windows washed by men on an outside hanging scaffold, the plans for the support of the scaffold over the parapet wall was not properly thought through. The bracing of the parapet covering was inadequate and the weight of men, and scaffold hanging over it crushed the parapet and caused water leakage. The entire parapet had to be recovered and reinforced at heavy expense.

If this section sounds like a catalogue of horrible examples, it is to emphasize the fact that none of these situations need have occurred. **One or two study meetings between the architect engineer and owner to go over the sensitive points in a plan and to ask each other "How is this likely to stand up under heavy use or is there enough capacity or can this be serviced properly?" will avert many future problems and possible ill will.**

1C. The Setting Up of a Maintenance Program

The architect engineer should be consulted by the owner and should be concerned with the setting up of a proper maintenance program. If the materials and methods of construction shown in his plans and described in his specifications have been properly used and the construction inspection properly carried out, there remains the matter of keeping everything well preserved and running smoothly.

Maintenance programs should be carefully thought through. Questions should be asked as to whether it is more expensive in some cases to allow a piece of equipment to run to destruction and then replace it or make it last longer by constant maintenance. Maintenance must be tailored to a building's size and the intended uses for its equipment. A small building with only a few men may have an entirely different program than a large building that can afford an expert in electrical maintenance, or boiler room equipment, or plumbing. In setting up the program the advice of the architect engineer should be sought. He is the one who chose the material and equipment and he should advise on its lasting and operating qualities. Manufacturers' recommendations should be studied and the interaction of equipment looked into to determine whether the failure of a small inexpensive part could cause a chain reaction and great damage. It may be better sometimes to spend more on maintenance of certain equipment than seems economical in order to avoid tenant annoyance.

When all these various factors have been studied and weighed a written program should be prepared and agreed to by all concerned.

It is hoped that this chapter will alert all who read it to the need for careful and thoughtful preplanning, the carrying through of the plan during construction, and, finally, maintaining what has been built. It will make a successful building, and a happy owner, and can add immeasurably to an architect engineer's reputation.

INDEX

A

Absorption liquid chillers, 140–142
Acoustic ceilings, 235, 244, 245
Acoustic decking, 254
Acoustical plaster, 235, 272
Air conditioning, 134–159
   control system, 153–154
   cooling plants, 137–142
      absorption liquid chillers, 140–142
      centrifugal compressor systems, 137–140
      location of, 144–145
      reciprocating units, 142
   determination of, 135–137
   double-duct system, 148–150
   equipment sizes, 140
   exhaust system, 151–152
   fan rooms, 146
   filters, 154–156
   high-pressure system, 146–147
   low-pressure system, 146–147
   perimeter window units, 150–151

Air conditioning (*cont.*)
   plans for, 31–32
   single-duct system, 147–148
   sound abatement, 152–153
   specifications, 137–139
   water treatment, 154–156
Air rights, 63
Airborne noise, 242
Alcoa Building (San Francisco), 81
All electric buildings, 122, 133, 156–159
Alley Theatre, 87
American Concrete Institute (ACI), 22
American Institute of Steel Construction (AISC), 22, 85, 86
American Iron and Steel Institute, 105
American Society of Hardware Consultants, 287–288
American Society of Heating, Refrigerating and Air Conditioning Engineers (ASHRAE), 153
American Society of Mechanical Engineers (ASME), 131, 134

INDEX

American Society for Testing Materials
(ASTM), 22, 52, 85, 90, 105,
122, 196, 197, 202, 204, 212,
262, 280, 284
American Standards Association Code,
189, 190
American Water Works Associations,
167
American Welding Society, 79, 90, 284
Apartment buildings, 7–9, 291
   floors, 225–226
   occupancy requirements in electrical
      works, 116
   plans for, 25–37
   scheduling construction, 314–315
Architect engineer:
   the contract, 39–42
      field inspection, 40–41
      general provisions, 39–40
      payment, 42
      shop drawings, 41–42
      types of, 40
   future duties and responsibilities,
      335–339
Architectural Graphic Standards, 76
Architectural Woodwork Institute, 306
Asphalt shingles, 251–252
Asphalt tile, 224, 225

B

Bar graph, 311–319
Bathrooms, 32
   ceramic tile, 277–280
   plans for, 32
   toilet facilities, 161–163, 279–280
      accessories, 280
      metal enclosures, 279–280
Bearing capacity, 52, 66
   of soil, 61
Beinecke Rare Book Library (Yale
University), 213–215
Bidding, 42–43
   finishing interior of business build-
      ings, 292
   general contractors, 42–43
   how solicited and assembled, 44
   subcontractors, 43, 44

Billets (steel) foundation, 79–80
Blackout of 1965, 120, 122–123
Blasting rock, 54–55
Brandeis University, 219, 221
Building codes:
   basic requirements, 18–20
   cities, 21–23
      fire protection, 22
      health regulations, 23
      safety regulations, 23
      structural requirements, 22–23
   electrical work, 20
   fire resistance, 18–19, 20, 22
   general purposes, 18
   health regulations, 20, 23
   heating, 20, 134
   light, 20
   plumbing, 20–21, 23, 163–169
   safety regulations, 18–19, 22, 23
   sanitation, 20
   small community, 20–21
   structural strength, 19–20, 22–23
   ventilation, 20
Building envelope, 34–37
   mechanical components, 35–36
   treatment of, 36–37
Business buildings:
   finishing the interior, 291–296
      bidding procedure, 292
      partitions, 292–296
      required flexibility, 292
   large, 6–7
      ceramic tile, 278–279
      floors, 227–230
      heating, 130–133
      scheduling construction, 315–319
      site planning, 10
      site selection, 5
      sound conditioning, 240–243
   millwork, 307–309
      doors, 308–309
      quality of material, 308
      workmanship, 309
   modular plan, 292
   planning interiors, 296–297
   plans for, 25–27
   small:
      ceramic tile, 278–279

Business buildings (*cont.*)
   electrical installation, 113–115
   floors, 225–226
   heating, 128–130
   roofing, 254–255
   scheduling construction, 314–315
   sound conditioning, 240–243
   suburban, 6

C

Cabinet work, 305–306, 307
Caissons (foundation), 66
Caniff, Charles E., 322–323
Carpeting, 230
Casualty insurance, 325
Ceilings, 231–235
   acoustic, 235, 244, 245
   composite, 232–233
   hung, 230, 231–232, 265, 269
   luminous, 233–234
   resiliently suspended, 245
   special-use, 234–235
Cellular steel floor, 105
Cement:
   Keene's, 272
   Portland, 273, 278, 279
   stainless, 207
Centrifugal compressor systems, 137–140
Ceramic tile, 224–225, 277–280
   large business buildings, 278–279
   private residence, 277–278
   small business buildings, 278–279
   special uses, 279
Chair (steel reinforcement), 90
Chesapeake Bay Bridge, 66
City:
   building codes, 21–23
      fire protection, 22
      health regulations, 23
      safety regulations, 23
      structural requirements, 22–23
   planning, 14–15
   zoning, 14–18
Clay tile, 252, 266

Columbia Broadcasting System Building (New York City), 92, 93
Composite ceilings, 232–233
Concrete:
   floors, 104–108
   frames, 87–96
      basic design, 88–89
      design systems, 91–94
      foundation, 91
      post-tensioned members, 94–96
      prestressed members, 94–96
      specification, 89–91
   roofing, 256–259
   safety protection, 330–331
   stairways, 249–250
   testing, 90
   walls, 198–200
      specifications, 200
   weather in placing, 91
Concrete blocks, 266
Contracts:
   architect engineer, 39–42
      field inspection, 40–41
      general provisions, 39–40
      payment, 42
      shop drawings, 41–42
      types of, 40
   construction, 44–48
      cost estimators, 48
      cost plus fixed fee, 45
      cost plus percentage, 45
      general conditions, 47–48
      guaranteed maximum plus fixed fee, 46–47
      lump sum, 44–45
      negotiated, 45–46
      scope, 47
Conveyors, 182–183
Corrugated steel floor, 105–106
Cost estimators, 48
Cost plus fixed fee contract, 45
Cost plus percentage contract, 45
Critical Path Method (CPM), 319–320
Curtain walls, 201–211
   examples of, 202–206
   glazing, 210–211
   prefabricated, 211
   specifications, 207–210

INDEX

D

Decibel (sound) 238, 240
Decking, roof, 254–255
Decorating, 273–275
  specifications, 275
Dewatering (excavation), 55–56
Doors:
  hollow metal, 285–286
  millwork, 306–309
Drilled in Caisson Company, 68
Drinking fountains, 165
Dry wall partitions, 295–296
Dumbwaiters, 182–183
Dyne, 238

E

Eiffel Tower (Paris), 82
Electric stairway, *see* Moving stairways
Electrical work:
  blackout of 1965, 120, 122–123
  building codes, 20
  installation, 111–123
    distribution determination, 111–119
    distribution system, 117–119
    emergency generators, 122–123
    equipment requirements, 116–117
    equipment selection, 120
    large business building, 115
    lighting, 121–122
    multidwelling building, 115
    occupancy requirements, 116
    power company distribution, 119–120
    private residence, 111–113
    requirement determination, 111–119
    small business building, 113–115
    specification, 115–116
  plans for, 30
Elevators, 176–180, 183–184
  braking system, 188
  cabs, 193
  door operation, 190–193

Elevators (*cont.*)
  freight, 178–179, 189
    equipment, 189
    specifications, 189
  material transportation by, 179–180
  need for, 337
  openings, 193
  passenger, 176–178, 183–193
    equipment, 184–193
    specifications, 184–193
  plans for, 32–33
  safety requirements, 176, 179, 189–190
  sidewalk, 179, 180
  special requirements, 183–184
  trim, 193
  worker transportation by, 179–180
Epoxy, 229
Equitable Life Building (Chicago), 301
Escalators (moving stairs), 181–182
  plans for, 32–33
Evaporation, latent heat of, 142
Excavation, 51–64
  dewatering, 55–56
  general, 54
  pier footings, 61–63
  pits, 61–63
  problems, 63–64
  rock, 54–55, 64
  safety protection, 330–331
  sheet piling, 57–61
  soils:
    bearing capacity of, 61
    soft, 54
    stabilization, 56
    underlying, 53
  test borings, 52–53
  test pits, 51–52
  underpinning, 57–61
  water bearing, 55–56, 63
Expansion joints, 260–261
Exterior walls, *see* Walls, exterior
Extruded shape, 202

F

Federal Bureau of Standards, 305

INDEX

Federal Housing Administration, 240
Fences for public protection, 326
Fire division, 285
Fire extinguishers, 331
Fire fighting systems, 171–173
    pumps, 171–172
    sprinklers, 32, 161, 172–173
    standpipes, 171–172
Fire insurance, 325
Fire protection, 32, 161
    building codes, 18–19, 20, 22
    during construction, 331–333
        fire extinguishers, 331
        fire watch, 332
        housekeeping, 332
        standpipes, 331
        vertical transportation, 332–333
    of steel, 85–87
Fire watch, 332
Fireproof buildings, 176, 248–250, 289
Fire-resistant, 74
First National Bank Building (Chicago), 82
Flake board, 308
Flashing, 252–253
    concrete structures, 256–259
    steel structures, 256–259
Floating floors, 244
Floors (structural), 101–108
    asphalt tile, 224
    carpeting for, 230
    ceramic tile, 225
    concrete, 104–108
    finished, 223–230
        apartment building, 225–226
        heavy-use, 226–227
        large office buildings, 227–230
        private residence, 223–225
        small office building, 225–226
        special-use, 226
    floating, 244
    forward planning, 338
    materials:
        description of, 102–104
        determination of, 101–102
    pedestal (raised), 229
    plans for, 29

Floors (cont.)
    quarry tile, 227–229
    reinforced concrete building, 106–108
        adaptability, 107–108
    resilient, 227
    rubber, 226
    structural steel building, 104–106
        cellular steel, 105
        corrugated steel, 105
        reinforced concrete, 104–105
        specification, 105–106
        steel deck, 106
    system determination, 101–102
    terrazzo, 227–229
    vinyl tile, 224
    waffle, concrete, 93
    wood parquet, 224
Foot-candles, 121
Footings, 64–69
    caissons, 66
    pier, 61–63
    piling, 66–70
    spread, 64–65
    stepped, 65
Ford Foundation Building (New York City), 213, 216–217, 303
Foundations, 64–71
    footings, 64–69
        caissons, 66
        pier, 61–63
        piling, 66–70
        spread, 64–65
        stepped, 65
    plans for, 29
    waterproofing, 70–71
Frame (structural), 73–99
    construction methods, 75–99
        basic design, 78, 88–89
        concrete, 87–96
        connections, 85
        design systems, 85, 91–94
        fire protection, 85–87
        foundation, 79–80, 91
        heavy wood, 96–99
        light steel, 76–77

INDEX

Frame (structural) (*cont.*)
    post-tensioned members, 94–96
    prestressed members, 94–96
    specification, 78–79, 89–91
    structural concrete, 87–96
    structural steel, 77–87
    variations, 80–85
    wall bearing, 76–77
    wood, 76, 96–99
    determining the type of, 73–75
        the architectural concept and, 75
        cost, 75
        labor availability, 74
        material availability, 74
        time of erection, 75
    plans for, 28–29
    for reinforced concrete stairs, 249
    testing, 79
Franki Pressure Injected Footing, 68, 69
Freight elevators, 178–179, 189
    equipment, 189
    specifications, 189
Freon (refrigeration), 137, 141
Frequencies, 238–239
Fuel, choice of, 133–134
Furring, 267

G

Gateway Center (Pittsburgh), 10, 11
Generators:
    emergency, 122–123
    hot water, 166
Glass, 210–211
    masonry wall and, 207–210
Glazing of a curtain wall, 210–211
Goggles, safety, 327
Grand Central Station (New York City), 63
Grillage, steel, 80
Guaranteed maximum plus fixed fee contract, 46–47
Gutters, 252–253
Gypsum block, 266
Gypsum lath, 270–271
Gypsum plaster, 271–272, 278

H

Handicapped, designing for, 321–323
Hardware finishing, 287–289
Harrison, Abromavitz, and Abbe, 83, 215, 220, 221
Hats for worker protection, 327
Health regulations in building codes, 20, 23
Heating, 125–134
    building codes, 20, 134
    fuel choice, 133–134
    large office building, 130–133
        distribution, 133
        plant choice, 130–133
        requirements, 130
    multiple dwelling, 128–130
        distribution, 129–130
        plant choice, 129
        requirements, 128–129
    plans for, 31–32
    private residence, 126–128
        requirement determination, 126
        specification, 128
        systems, 126–128
    radiant, 127
    small business building, 128–130
    solar, 127
Height restrictions, 15–18
    in business areas, 16–17
    in residential areas, 16
    in transitional areas, 17–18
Hung ceilings, 230, 231–232, 265, 269

I

Industrial wastes, 171
Inspection during construction, 335–337
Insurance, 325
Intensity of sound, 238–239
Iron work, 261–263
    specifications, 262

J

Jewish Chapel (Brandeis University), 221

INDEX

John Hancock Building (Chicago), 81, 82, 184
Joints, expansion, 260–261

K

Keene's cement plaster, 272
Kevin Roche, John Dinkeloo, 216
Kitchens:
    ceramic tile for, 277–278
    plans for, 32

L

Laboratories, 25, 27–28, 161
    electrical installation, 114
    occupancy of, 291
    plans for, 25–37
    plumbing, 170
    site planning, 10
    site selection, 5–6
Landscaping, 299–303
    zoning in, 300
Lathing, 267–271
    expanded metal, 268–270
    gypsum, 270–271
    specification for, 269
Le Corbusier, 198, 199
Leaders (roofing), 252–253
L'Enfant Plaza (Washington, D.C.), 93, 94, 108
Letter of Invitation (bidding), 44
Light leaks, 122
Light spill, 121
Lighting, 20
    building codes, 20
    electrical installation, 121–122
    systems of, 93, 115
Lime putty, 272
Line drilling, 55
Live load, 22
Lobbies, 184
Luminous ceilings, 233–234
Lump sum contract, 44–45

M

Maintenance, 338–339
    plans for, 33–34

Masking noise, 239
Masonry:
    interior, 265–267
        code requirements, 265–266
        materials, 266
        specifications, 267
    walls, 196–198
        glass and, 207–210
        specifications, 207–210
Medical buildings:
    access to, 322
    electrical installation, 114
    occupancy of, 291
    plans for, 25–37
    plumbing, 170
    site planning, 10
    site selection, 5–6
Metal:
    hollow metal work, 285–286
    ornamental, 283–285
    partitions, 293–295
Metal lathing, 268–270
Metal Roof Deck Technical Institute, 105
Metropolitan Opera (New York City), 247
Microns, 155
Millimeters, 155
Millwork, 305–309
    business building, 307–309
        doors, 308–309
        quality of material, 308
        workmanship, 309
    private residence, 305–307
        doors, 306–307, 308
        paneling, 307
        trim, 307
        windows, 306
Mockups, 211–212
Morse College (Yale), 196, 213
Moving stairways (escalators), 181–182
    plans for, 32–33
Multiple dwellings (apartment houses):
    building codes, 20–21
    client needs, 27
    electrical installation, 115
    heating, 128–130
        distribution, 129–130

INDEX

Multiple dwellings (*cont.*)
    plant choice, 129
        requirements, 128–129
  plans for, 25–37
  roofing, 254–255
  site planning, 7–9
  site selection, 4–5
  sound conditioning, 239–240
Museum of Modern Art (New York City), 301–302

N

National Association of Architectural Metal Manufacturers (NAAMM), 212
National Board of Fire Underwriters, 22, 285
National Bureau of Standards, 155, 156, 204, 240, 278
National Electrical Code, 20, 23
National Electrical Manufacturers Association, 120
National Fire Underwriters' Code, 20
National Hardwood Lumber Association, 305
Negotiated contract, 45–46
Neoprene, 210
Noise, 336
  airborne, 242
  background, 297
  criterion (NC), 239, 245
  human, 241–242
  inside, 241
  machine, 242–243
  masking, 239
  outside, 241
  pollution, 237
  vibrations, 242–244
Nylon carpeting, 230

O

Openings (construction), protection of, 327–328
Ornamental metal, 283–285
Ornamental stairways, 251

P

Painting, 273–275
  specifications, 275
Paneling, 307
Parging brick, 197
Parklabrea (Los Angeles), 9
Partitions in business buildings, 292–296
  choices, 293
  dry wall, 295–296
  metal, 293–295
  soundproofing, 294
Party wall, 330
Passenger elevators, 176–178, 183–193
  equipment, 184–193
  specifications, 184–193
Pedestal (raised) floors, 229
Pennsylvania Station (New York City), 63
Perlite, 254, 273
pH factor in water, 155, 168
Philharmonic Hall (New York City), 213, 215
Phoenix Mutual Life Building (Hartford), 219, 220
Pier footings, 61–63
Piling, 66–70
  bearing capacity, 66
  sheet, 57–61
  soldier, 60–61
Piping, plans for, 34
Pits, 61–63
  test, 51–52
Plans, 25–37
  the architectural concept, 34–37
  building costs, 28
  building envelope, 34–37
    mechanical components, 35–36
    treatment of, 36–37
  client needs, 25–28
    multifamily dwellings, 27
    office buildings, 25–27
    private residence, 25
    special-purpose buildings, 27–28
  the design, 28–34
    structural, 28–29

Plans (cont.)
  electrical system, 30
  elevators, 32–33
  floor system, 29
  the foundation, 29
  the frame, 28–29
  laboratories, 25–37
  maintenance considerations, 33–34
  mechanical design, 31–32
    air conditioning, 31–32
    heating, 31–32
    plumbing, 32
    vertical transportation, 32–33
  medical buildings, 25–37
  private residence, 25–37
  site, 7–11
  stairways, 32–33
Plastering, 271–273
  acoustical, 235, 272
  gypsum, 271–272, 278
  Keene's cement, 272
  Portland cement, 273, 278, 279
  specifications, 273
  vermiculite, 272
Plastic, 85, 122
Plastic lenses, 122
Plumbing:
  building codes, 20–21, 23, 163–169
  installation, 161–171
    accessories, 168–169
    building requirements, 163–169
    code requirements, 161–163
    distribution system, 169–171
    fixtures, 168–169
    industrial wastes, 171
    materials, 166
    methods of, 166
    process water, 171
    sanitary facilities, 161–167
    specification, 166
    storm drainage system, 166
    system design, 165
    tests, 169
    water supply, 163–165, 167–168
  plans for, 32
Pollution, 163, 171
  noise, 237
Polyester (plastic), 229

Polyethylene (plastic), 224
Polystyrene plastic (lens), 122
Polyvinyl acetate base (PVA) (paint), 274
Polyvinyl chloride base (PVC) (paint), 274
Porcelain Enamel Institute, 284
Portland cement, 273, 278, 279
Portland Cement Association (PCA), 22
Portland cement plaster, 273
Post-tensioned members, 94–96
Potable water, 164
Power company distribution, 119–120
Prefabricated walls, 201, 211
  curtain, 211
Prestressed members, 94–96
Private residence, see Residence, private
Process water, 171
Projectile tools, 329–330
Psychrometer, sling, 136
Public protection, 326–327
  fences, 326
  general measures, 326–327
  sidewalk sheds, 326
Pumps, 55, 56, 164, 166, 167
  in fire fighting systems, 171–172
  heat, 127–128

R

Radiant heating, 127
Residence, private:
  ceramic tile, 277–278
  electrical installation, 111–113
  flashing, 252–253
  floors, 223–225
  gutters, 252–253
  heating, 126–128
    requirement determination, 126
    specification, 128
    systems, 126–128
  leaders, 252–253
  millwork, 305–307
    doors, 306–307, 308
    paneling, 307

INDEX

Residence, private (cont.)
    trim, 307
    windows, 306
  plans for, 25–37
  roofing, 251–253
  scheduling construction, 313–314
  site planning, 7
  site selection for, 3–4
  sound conditioning, 239–240
  stairways, 247–248
Resin, 229
Robertson Q air floors, 149
Rock excavation, 54–55, 64
Rockefeller Center (New York City), 11, 301, 302
Ronchamp Chapel (France), 199
Roofing, 251–261
  concrete structures, 256–259
  expansion joints, 260–261
  factories, 259–260
  institutional building, 255–256
  iron work, 261–263
  monumental building, 255–256
  multiple dwelling, 254–255
  private residence, 251–253
  small office building, 254–255
  steel structures, 256–259
  warehouses, 259–260
Rubber floors, 226
Rubbish removal, 328
Rudolph, Paul, 198

S

Saarinen, Eero, 92
Safety belts, 327
Safety during construction, 325–333
  concrete construction, 330–331
  excavation, 330–331
  fire protection, 331–333
    fire extinguishers, 331
    fire watch, 332
    housekeeping, 332
    standpipes, 331
    vertical transportation, 332–333
  meetings on, 325–326
  public protection, 326–327
    fences, 326
    general measures, 326–327

Safety during construction (cont.)
    sidewalk sheds, 326
  steel construction, 330–331
  worker protection, 327–330
    hats, 327
    openings, 327–328
    projectile tools, 329–330
    rubbish removal, 328
    safety belts, 327
    scaffolds, 329
    vertical transportation, 328–329
Safety regulations:
  building codes, 18–19, 22, 23
  elevators, 176, 179, 189–190
Sand float finish (plaster), 273
Sanitation, 278
  building codes, 20
  facilities, 161–167
  the waste system, 166
Scaffolds, 329
Scheduling construction, 311–320
  apartment building, 314–315
  bar graph, 311–319
  Critical Path Method (CPM), 319–320
  importance, 311–319
  large business building, 315–319
  private residence, 313–314
  small business, 314–315
Scope contract, 47
Screeding, 228
Seismographs, 55
  in inspection of adjoining property, 71
Sewage ejectors, 166, 167
Sheet piling, 57–61
Shingles:
  asphalt, 251–252
  wood, 251
Shop drawings, 41–42
Siamese connection, 172
Sidewalk elevators, 179, 180
Sidewalk sheds, 326
Site adjoining property, 71
Site planning, 7–11
Site selection, 3–7
Skidmore, Owings and Merrill, 213, 218, 219
Sky lobbies, 184

INDEX

Slate, 252
Slump testing, 90
Soils:
   bearing capacity of, 61
   soft, 54
   stabilization, 56
   underlying, 53
Solar heating, 127
Soldier piling, 60–61
Sound:
   abatement, 152–153
   character of, 238–239
   insulation, 338
   intensity of, 238–239
Sound conditioning, 237–245
   character of sound, 238–239
   designing for, 239–245
      human noise, 241–242
      inside noise, 241
      large office buildings, 240–243
      machine noise, 242–243
      multidwelling, 239–240
      outside noise, 241
      private residence, 239–240
      small office buildings, 240–243
      special problems, 243–245
   requirements, 237–239, 240
Soundproofing of partitions, 294
Specifications:
   air conditioning, 137–139
   decorating, 275
   electrical installation, 115–116
   elevators:
      freight, 189
      passenger, 184–193
   frames, 78–79, 89–91
   heating, 128
   interior masonry, 267
   iron work, 262
   lathing, 269
   metal partitions, 294
   painting, 275
   plastering, 273
   plumbing, 266
   steel floors, 105–106
   steel frames, 78–79
   vertical transportation, 184–193
   walls, 207–210
      concrete, 200

Specifications (*cont.*)
   curtain, 207–210
   writing of, 37
Spencer, White and Prentis Company, 68
Spline, 231
Spread footings, 64–65
Sprinkler systems, 32, 161, 172–173
Stairways, 175–176, 247–251
   concrete, 249–250
   in fireproof buildings, 248–250
   moving, 181–182
      plans for, 32–33
   ornamental, 251
   plans for, 32–33
   private residence, 247–248
   steel pan, 248
Standpipes, 161, 171–172, 331
Steel:
   frames, 77–87
      basic design, 78
      connections, 85
      design systems, 85
      fire protection, 85–87
      foundation, 79–80
      light, 77
      specification, 78–79
      variations, 80–85
      walls, 200–201
Steel structures:
   flashing, 256–259
   floors, 104–106
      cellular, 105
      concrete, 104–105
      corrugated, 105
      deck, 106
      specification for, 105–106
   roofing, 256–259
   safety protection, 330–331
Stepped footing, 65
Stiles College (Yale), 196, 213
Storm drainage system, 166
Subcontractors, bidding, 43, 44
Sump pumps, 166, 167

T

Taxpayer buildings, 25–26
Terrazzo floors, 227–229

351

INDEX

Test borings, 52–53
Test cylinder, 90
Test pits, 51–52
Testing:
  concrete, 90
  exterior walls, 211–212
  frames, 79
  plumbing, 169
  slump, 90
  wind loading, 212
Testing agency, 90
Thornley, J. H., 68
Tile:
  asphalt, 224, 225
  ceramic, 224–225, 277–280
    large business buildings, 278–279
    private residence, 277–278
    small business buildings, 278–279
    special uses, 279
  clay, 252, 266
  quarry, 227–229, 278, 279
  vinyl, 224, 225
Toilet facilities, 161–163, 279–280
  accessories, 280
  metal enclosures, 279–280
Transformer vault, 113
Transformers, 113
Trim, 307
Two-hour walls, 22

U

Underground streams, 55
Underpinning, 57–61
U.S. Post Office Department, 183
United States Steel Building, 82–84
University Heights (New York), 8
University of Illinois Assembly Hall, 88, 89

V

Ventilation, 134–159
  building codes, 20
  required, 135
Vermiculite, 254

Vermiculite plaster, 272
Vertical transportation, 175–193; *see also* types of vertical transportation
  equipment, 184–193
  for firemen, 332–333
  plans for, 32–33
  requirement determination, 175–184
    conveyors, 182–183
    dumbwaiters, 182–183
    elevators, 176–180, 183–184
    moving stairways, 181–182
    stairways, 175–176
  specifications, 184–193
  worker protection, 328–329
Vibrations, 242–244
Vinyl tile, 224, 225
Virginia National Bank Building (Norfolk), 213, 217–218, 219, 220

W

Waffle floor (concrete), 93
Wallboard, 271
Walls, *see also* Partitions of business buildings:
  bearing, 76–77
  cavity, 197
  concrete, 198–200
    specifications, 200
  curtain, 201–211
    examples of, 202–206
    glazing, 210–211
    prefabricated, 211
    specifications, 207–210
  exterior, 195–221
    design determination, 195–211
    mockups, 211–212
    performance standards, 211–212
    specially designed, 212–221
    tests, 211–212
  masonry, 75, 77, 196–198
    glass and, 207–210
    specifications, 207–210
  party, 330
  perimeter, 62

INDEX

Walls (*cont.*)
   prefabricated, 201, 211
      curtain, 211
   steel, 200–201
   two-hour, 22
Water:
   in air conditioning, 154–156
   chilled, 165
   drinking, 161, 164–165
   evaporation of, 142
   excavation and, 55–56, 63
   hard, 34
   pH factor, 155, 168
   potable, 164
   process, 171
   soft, 34, 167
   systems:
      cold, 167
      hot, 168
Water pollution, 163
Water supply, 163–165, 167–168

Waterproofing:
   foundations, 70–71
   roof, 254–255
Weidlinger, Paul, 92
Weiskopf & Pickworth, 94, 95
Wellpoints, 55–56
Wind loading tests, 212
Windows, 77, 306
Wood Characteristics Table, 306
Wood frames, 76
   heavy, 96–99
Wood parquet, 224
Wood shingles, 251
Wool carpeting, 230
Worker elevators, 179–180
Worker protection, 327–330
   hats, 327
   openings, 327–328
   projectile tools, 329–330
   rubbish removal, 328
   safety belts, 327